Realm of the Long Eyes

A Brief History of Kitt Peak National Observatory

Front Cover Illustration: Standing 187 feet high, the building for the 158-inch telescope dominates the site at Kitt Peak National Observatory. Containing thirty times the floor space of an average house, the interior of the building is divided into offices, workshops, darkrooms, sleeping quarters, and storage areas. *(Photograph by the author.)*

Frontispiece: Kitt Peak National Observatory showing the 60-foot site survey tower as it appeared atop Kitt Peak in the Quinlan Mountains of southwestern Arizona during 1957. The triple shelled triangular structure housed the six-inch site survey telescope and various instruments used to measure wind velocity, temperature, and humidity. During the initial phases of the site survey, staff members visited the tower once a month to collect the data and maintain the recording and power equipment located in the base of the tower. Note the caterpillar tractor used to haul equipment up the steep grades of the primitive site survey road.

AN INDEPENDENT, UNOFFICIAL HISTORY OF KITT PEAK NATIONAL OBSERVATORY. THE ASSOCIATION OF UNIVERSITIES FOR RESEARCH IN ASTRONOMY (AURA) IS NOT RESPONSIBLE FOR ITS CONTENTS.

This book is dedicated to the men and women who gave unselfishly of their time and talents in the establishment of Kitt Peak National Observatory, and to those individuals through whose efforts Kitt Peak continues to rank as a truly remarkable research institution.

Realm of the Long Eyes

A Brief History of Kitt Peak National Observatory

by
James E. Kloeppel

Published by Univelt, Inc., San Diego, California

Foreword

The American public watched with fascination as the 200-inch glass giant of Mount Palomar took shape. Books were written about the crowning achievements of George Ellery Hale as he built the world's largest telescope. Historical accounts were written showing the telescope's progress as construction took place high on an isolated mountain top.

That was half a century ago. Although there has been a plethora of new telescopes built throughout the world since the time of George Ellery Hale, little has been written for the public about their construction or development. This book was written in an attempt to bridge a part of that gulf.

Kitt Peak National Observatory, nestled among the pinon pines in the Quinlan Mountains of southwestern Arizona, houses the largest concentration of optical telescopes and astronomical instrumentation to be found anywhere in the world. Each year, hundreds of astronomers, out of a field of only a few thousand, come from across the globe to use the Kitt Peak telescopes as they attempt to unlock the mysteries of the universe. Additionally, nearly a hundred thousand tourists visit Kitt Peak annually. Magnificent views across hundreds of miles of Arizona's rugged landscape, coupled with equally impressive views of the large telescopes poised for the night's work, create memories not soon to be forgotten.

This book is a brief history covering the development of Kitt Peak National Observatory. I have attempted to outline briefly how the challenge to create a national observing center was met, how Kitt Peak came to be chosen as the site for the new National Observatory, and how the site was developed from a primitive mountain top to a modern astronomical research facility.

vii

Like many of the old mining towns which dot the American landscape, Kitt Peak National Observatory clings precariously to the top of a mountain. Unlike the mining towns — which gather the riches of the earth — Kitt Peak is gathering the treasures of the universe. Astronomers at the observatory are studying the far-flung stars of space. Each tiny photon of light which strikes their instruments has a meaning for the Kitt Peak astronomers. With meaning comes understanding, and with understanding comes appreciation. A better appreciation of the universe in which we live will lead to a better understanding and appreciation of ourselves, and of how we fit into the cosmic puzzle. I commend the scientific and support staff of Kitt Peak National Observatory. And I commend the American public who made the observatory possible.

In a historical work of this nature it is impossible to give credit to all the people who helped by answering questions or providing much appreciated tidbits of information. However, a few individuals come to mind without whose help this book could never have been written. In particular, I would like to thank Bill Daggett and J. C. Golson for their support and encouragement. For their help with the historical information I wish to thank Dr. Leo Goldberg, Dr. Helmut Abt, and Dr. Dave Crawford. For providing much needed information on casting the 84-inch and 158-inch glass blanks, I would like to thank Wm. C. Lewis of Corning and Robert J. Sommer of General Electric. I would also like to express my gratitude to Agnes Paulsen for her help in acquiring the Kitt Peak photographs. Additionally, I would like to thank my special friends Dean Ketelsen, Dan Brodzik, George Will, and Dr. Dave Monet for making this project so enjoyable. Lastly, I would like to thank my wife Darlene, who spent many lonely nights while this text was being prepared, for her support.

James E. Kloeppel

Contents

*There is no richer field of science
opened to the exploration of man in search
of knowledge than astronomical observation;
nor is there, in the opinion of this committee, any duty more impressively incum-
bent upon all human governments than that
of furnishing means and facilities and
rewards to those who devote the labors of
their lives to the indefatigable industry,
the unceasing vigilance, and the bright
intelligence indispensable to success in
these pursuits.*

John Quincy Adams, 1842

Vision of a National Observatory

*I think what this country needs is a
truly National Observatory to which every
astronomer with ability and a first class
problem can come...on leave from his
university.... We ought not to be as
much concerned with the possibility of
doing useful work as with the real need
to do great work....*[1]

Dr. Leo Goldberg

The date was August 31, 1953. Thirty-five astronomers had
gathered together at Lowell Observatory in Flagstaff, Arizona,
to discuss America's growing need for more modern
astronomical research facilities. Acting upon a proposal put forth
by the University of Arizona, Indiana University, and Ohio State
University, a then very young National Science Foundation
(NSF) had sponsored this "Astronomical Photoelectric Conference" in order that astronomers from across the country
might discuss, among other things, the idea of a jointly funded
and operated telescope for photoelectric research.*

Astronomers had several reasons for desiring such a
telescope, one of which was that the existing medium and large
telescopes could not adequately serve the astronomical community. These telescopes were privately financed, and were
available primarily to astronomers from the parent institution.

---

*Photoelectric research (photometry) deals with the measurement of the intensity
of starlight.

1

While some observing time was allocated to astronomers from other institutions — Mount Wilson in particular gave a generous portion of available time to visiting astronomers — there were far too many astronomers for the available telescope time. Also, the granting of such time was proving to be a drain on the parent institutions; not only were the staff astronomers sacrificing precious observing time, but they had to be available for instructing the visitors on the use of the instruments, and for providing assistance when problems developed. Additionally, with the continued growth of the parent institutions' astronomy departments, more time was needed for their own staffs and therefore less time would be available for visitors.

Astronomical telescopes, instrumentation, and the domes to house them are very expensive. American colleges and small universities simply did not possess the necessary funds to construct major observing instruments, so the bulk of the astronomical community could not easily collect the data they needed and American astronomy was suffering because of it.

Dr. Bart J. Bok brought forth another problem associated with existing telescopes:

> *When we examine the sites selected for observatories built in the past seventy years ... we find that with a very few exceptions, the present observatories are located in the wrong place. Not only are many of our most excellent instruments located in poor climates, but time and again one finds them in or near cities, large towns or industrial centers which generate smoke and dust and produce undue sky illumination ...* [2]

Due to deteriorating observing conditions these telescopes were losing their effectiveness in attacking frontier problems in astronomy. Recent advancements in photoelectric technology, coupled with the need to attack modern astronomical problems demanded more modern telescopes, located in more remote sites throughout the country.

But the concept of a jointly operated telescope utilizing state-of-the-art photoelectric devices rapidly came under fire at the conference. Those present disagreed on the size and type of telescope that should be constructed. Others were concerned about where the telescope should be located, how observing time could be equitably shared, and who would be responsible for building and maintaining the telescope and instruments. Some institutions which were represented at the conference were eager to venture forth with the project, others were ambivalent, and still others flatly refused to participate.

Complicating the situation even more, one of the astronomers attending the conference had an even bigger idea. What Dr. Leo Goldberg envisioned was a much larger facility, with several telescopes and major instrumentation, to be used not just by photometric observers, but by all astronomers. Dr. Goldberg stated:

> As I look around the room here, I count about twenty very competent researchers who do not now have access to first class instruments, to say nothing of a large telescope. If you went around the country and included other areas in astronomy you could conservatively get that number up to fifty ... I think what this country needs is a truly National Observatory to which every astronomer with ability and a first class problem can come ... on leave from his university.... We ought not to be as much concerned with the possibility of doing useful work as with the real need to do great work....*

Dr. Goldberg went on to explain that those attending the conference comprised only a narrow segment of the astronomical community. Other groups, stellar spectroscopists,† for example, could present equally convincing arguments for acquiring observing time on joint telescopes. Whereas the older telescopes were individually funded and open only to astronomers from the particular institutions, a National Observatory, Dr. Goldberg maintained, would be different. The observatory would be a national center for the entire scientific community, built by Federal funds, and available for use by all astronomers, regardless of the person's affiliation.

---

*See Reference Note 1

†Stellar spectroscopists are astronomers who, by spreading the light from a star into a spectrum, can determine such physical characteristics as a star's motion or its chemical composition.

While some astronomers were hesitant, many hoped Dr. Goldberg's suggestion would take root and grow under the guidance of the National Science Foundation. But, could the embryonic NSF support such an expensive, major undertaking? The 35 astronomers passed a resolution recommending to the NSF that a special committee be appointed to study the problems associated with establishing a National Observatory.

When the National Science Foundation Advisory Panel for Astronomy met the following January, the idea of a general survey of telescopes was indeed endorsed. As a result, a special advisory panel consisting of Dr. Robert R. McMath as Chairman (McMath-Hulbert Observatory), Dr. Ira S. Bowen (Mount Wilson and Palomar Observatory), Dr. Bengt Stromgren (Yerkes Observatory), Dr. Otto Struve (Leuschner Observatory), Dr. A.E. Whitford (Washburn Observatory), and Dr. Leo Goldberg (University of Michigan) was appointed by the NSF. The panel was to study sites, instrumentation, costs, and other factors relating to the establishment of a National Optical Observatory.

While the panel gave considerable thought to the role that new telescopes and advanced instrumentation might play in the growth of American astronomy, it was recognized that new telescopes and instruments alone would not be enough; one of the prime needs was for more and better astronomers. Although the research to be done at the new observatory would provide ample justification for building it, of equal importance would be its educational functions. In an after-dinner talk to the American Astronomical Society, Dr. Whitford said:

> *Such a national observatory center would lessen the reluctance of young astronomers to accept posts in institutions where opportunities have thus far been limited to teaching. By opening a path to continued professional development for a young astronomer, the new observatory could contribute very materially to raising the general level of staff competence in the colleges and small universities. This could result in an increase in the number of gifted students ... who would be influenced toward an astronomical career.*[3]

After studying the needs of American astronomers, the panel recommended that the National Science Foundation support construction of a National Astronomical Observatory. The panel advocated immediate construction of a 36-inch stellar telescope, an 80-inch stellar telescope, and what would be the world's largest solar telescope.

Site Selection

In the history of American astronomy, never was a site for an optical astronomical observatory selected with such meticulous care.[4]

Trudy Bell

The primary task facing the panel was to initiate a comprehensive survey of possible sites to determine an ideal location for the observatory. Dr. Aden B. Meinel (Yerkes Observatory), assisted by Dr. Helmut Abt (McDonald Observatory), was selected to conduct the site survey, an exhausting search that would take three years as the number of sites was reduced to the final choice.

In selecting the site, the primary considerations were an atmosphere of the highest possible transparency, excellent seeing, and a high percentage of clear skies.* These requirements called for the general region of the Southwest, preferably in a climatic zone with a weather pattern "out of phase" with that of the Pacific Coast, where so many of the large

◄— **Rising above the 7400-foot elevation of Hualapai Mountain stands the 60-foot site survey tower. This tower, along with similar units erected at other sites, was used to analyze the site's potential for a National Observatory.** *(Photograph by J. C. Golson, 1957)*

---

*"Transparency" is a term astronomers use to describe how free the atmosphere is from smoke, dust, and water vapor; while "seeing" refers to the stability of the atmosphere. Stars appear as sharp points of light when the seeing is good, but appear blurred and distorted when the seeing is poor.

The Pinal Mountains form a magnificent backdrop at Cutter Airfield near Globe, Arizona. Helmut Abt flew over 2000 miles during the site survey in this two-place Cessna 140 aircraft piloted by J.O. Casparis. Photographs of potential sites were taken from the airplane for the National Astronomical Observatory by Dr. Abt. *(Photograph by Dr. Abt, May 8, 1955)*

Site survey staff members Harold Thompson (in photograph) and Helmut Abt (the photographer) pause near an abandoned adobe hut while studying the region surrounding the base of Kitt Peak on March 13, 1956.

8

telescopes were situated.[5] The site survey area included the six western states of California, Arizona, New Mexico, Nevada, Utah, and Texas.

The initial selection of areas of interest was made on the basis of criteria that could be easily evaluated by visual inspection. Ground cover (necessary for good transparency and seeing conditions) at the site and over the prevailing air trajectory approach was a major factor. The site's ability to be developed was also of great importance, as the arrangement of the site area and its total developable area would be reflected in the costs of construction. Another factor in the site evaluation was accessibility, not only the means of access to the region but the location of the nearest towns and schools.

In the history of American astronomy, never was a site for an optical astronomical observatory selected with such meticulous care. Requirements for the ideal site were tough and exacting; the site must frequently be clear, with dry stable air; must be high enough to be above the dust, fog, and smoke of lowlands, but low enough to avoid the major weather patterns of high mountain regions; and must have a dark sky but, at the same time, must also be located near a supporting city with a major university. *

The first phase of the site survey began in 1955 by examining photographs taken by Viking rockets launched from White Sands Missile Range in New Mexico. The locations thus selected were further examined by Dr. Abt who flew over 2000 miles studying and photographing each potential site. This aerial reconnaissance narrowed the search to 150 sites, which were then examined on the ground. Nearly 4000 more miles were covered by Dr. Abt and Dr. Meinel as they explored the sites by jeep, pack horse, and foot.[6]

By 1956, on the basis of the ground examination results, all but five of the possible sites had been rejected. The remaining sites were: Chevalon Butte, southwest of Winslow, Arizona; Summit Mountain, south of Williams, Arizona; Hualapai Mountain, southeast of Kingman, Arizona; Kitt Peak, southwest of Tucson,

---

*See Reference Note 4.

9

Arizona; and Junipero Serra Peak, near Monterey in California. During the next year Chevalon Butte and Summit Mountain would be eliminated, and a new possibility, Slate Mountain, northwest of Flagstaff, Arizona would take their place.

To further evaluate these potential locations, special testing equipment was designed, built, and hauled (sometimes with great difficulty) to the various sites. Instruments to measure such site characteristics as transparency of the sky, wind velocity, relative humidity, and temperature fluctuations were installed in triangular steel towers which stood sixty feet tall. Each tower was constructed of three concentric but independent shells; the outer two shells were designed to protect the inner instrument tower from windshake.

Because no road existed to the summit of Kitt Peak, the site survey crew was forced to use pack horses to study the mountain in detail. This photograph by Helmut Abt shows Harold Thompson and a Papago guide leaving the San Vincente area near the base of the mountain for an assault on Kitt Peak on March 14, 1956.

10

An unidentified worker preparing to hoist the six-inch site survey telescope to the top of the 60-foot site survey tower atop Kitt Peak. The telescope and associated photoelectric equipment were capable of automatically monitoring the light of Polaris to determine the site's suitability for a National Observatory. *(Photograph, February 6, 1957)*

Despite the site survey tower's unique structural design to prevent vibrations from being transmitted to the telescope, windshake proved to be a problem at the Kitt Peak site. A shorter, ten-foot tower was constructed and hauled up the mountain. The six-inch automatic telescope was transferred to the shorter structure and experiments to determine the site's suitability were resumed. In this photograph workers are preparing to unload the tower and set it upright on its prepared foundation.

An ingenious method of automatically monitoring the seeing conditions at the sites was also devised. A six-inch telescope, mounted in the very top of each tower, was used to observe photoelectrically the light of the north star, Polaris. Starlight entering the telescope slowly traversed a grid of black (opaque) and white (transparent) lines. The light was then analyzed electronically by a photomultiplier tube whose output was permanently recorded on a slowly rotating strip chart recorder. When the seeing was good, as the starlight traveled across the grid the light level would change distinctly. During poor seeing, the contrast would diminish. Staff members visited the sites monthly to collect the data, and maintain the recording and power equipment located in the base of the towers.

Experience with the sophisticated automatic equipment under primitive operating conditions soon indicated that the Hualapai

Mountain site and the Kitt Peak site were the two best possible locations for the National Observatory. A decision was made to man the two sites in order to conduct standardized seeing and sky transparency tests. Therefore, in March of 1957 site survey member Claude Knuckles moved into a small housing trailer atop Kitt Peak and began visual and photoelectric observations with a small six-inch telescope. Another staff member, J.C. Golson, moved to the Hualapai site to perform similar duties. Soon two 16-inch reflector telescopes, built by Phemco in Phoenix, Arizona, were mounted on trailers and equipped with photoelectric photometers. One telescope went to Golson at Hualapai and was used between September 8, 1957, and July 8, 1958. The other telescope went to Knuckles at Kitt Peak where it was used between August 8, 1957, and October 6, 1958. The final decision between the two sites was made primarily on the basis of results obtained with the 16-inch telescopes.

Nestled among the ponderosa pine and manzanita brush which cover Hualapai Mountain stands a lonely housing trailer. J. C. Golson took this photograph in 1957 while he was based at the site during the site survey. Note the portable generator at the rear of the trailer.

13

The housing trailer and an assortment of equipment used on Kitt Peak during the site survey. In the center appears a six-inch refractor telescope initially used on the mountain. A portable generator, visible to the far left, was used to supply power to this remote site. *(Photograph, 1957)*

Claude Knuckles (at left) and Gary Chapman standing within the specially designed trailer which housed the 16-inch site survey telescope. This instrument was used on Kitt Peak from August 8, 1957, until October 6, 1958. Standardized tests were conducted with this telescope on Kitt Peak and a similar telescope on Hualapai Mountain.

14

Finally, after three years of searching and conducting exhaustive tests, the results were announced in March of 1958: Kitt Peak was clearly the best choice. In the *Final Report on the Site Selection Survey for the National Astronomical Observatory,* issued by Dr. Meinel, the relative merits of Kitt Peak and Hualapai Mountain were listed. In rainfall, sky transparency, and darkness of the night sky, the two mountain areas were about equal; and a water-supply problem existed at both sites. There were somewhat clearer nights on Hualapai, and the problems of road access, bringing in utilities, and land procurement would have been less than for Kitt Peak.

However, the latter was rated superior on eleven counts: good seeing, low wind velocity, temperature stability, microthermal stability, upper-air trajectories, absence of airplane vapor trails, developable area on the mountaintop, more southerly latitude, less interference from city lights, nearness of general support facilities, and proximity to an academic institution.[7]

Rising from the Sonora Desert plain to an altitude of 6,875 feet, the Kitt Peak land mass dominates its sister mountains in the Quinlan Range, as well as the lesser hills which stretch to the horizon well over a hundred miles distant. The Quinlans are a northerly continuation of the Baboquivari Mountains which extend from the Mexican border to the northern foothills of Kitt Peak.

Geologically, Kitt Peak is a fault-blocked, Pre-Cambrian granite mountain of little mineral significance. Its lower levels are covered with typical desert foliage including the familiar palo verde and mesquite trees, as well as several common varieties of cactus. With sufficient rainfall in a given year, lush range grasses appear, and the entire mountain presents a rich, verdant picture. Above the 5,000 foot level, thickening brush gives way to live oak trees, pinon pine, and an occasional small locust and walnut. Approximately eighty-five acres of the summit area are gently rolling grass lands interspersed with groves of native trees and outcroppings of rocks.[8]

The mountain was named by George J. Roskruge, who had arrived in the Arizona Territory in 1874 and became Pima County Surveyor. Among the many places Roskruge named in southern Arizona was Kitt Peak, which he named in honor of his sister Philippa (Roskruge) Kitt, who came to Arizona at his urging and died in 1900. The name Kitt Peak was finally made official by the United States Geographic Board in 1930. For many years it was spelled erroneously as "Kits" or "Kitts" Peak.[9]

The Sacred Mountain
of the Papago Indians

Ioligam, the Sacred Mountain in tribal legend, was a favorite dwelling place of I'itoi, the Papago Elder Brother. It was also a place cherished by the tribe for recreational visits. The Papagos would go up the mountain where the gods lived, enjoy its serenity and partake of its exotic mountain foods. When the time came for the return to the village, they would remember the generosity of I'itoi who had created the wild things, as they left their offering as a token of gratitude for the gifts that Elder Brother had bestowed upon them.

(Adapted from an unpublished paper by Geoffrey Anderson Clark) [10]

For hundreds of years this fairly steep mountain had been known by another name. Local inhabitants, the Papago Indians, had called the mountain "Ioligam" which meant mountain of manzanita wood. Historically, the Papago had been peace-loving Indians who made their livelihood by tilling the soil. The name Papago literally means "bean grower".*

The Papago have occupied the desert region surrounding Ioligam since at least the year 1539, when Spanish explorers searching for the legendary "Seven Cities of Cibola" encoun-

---

*See Reference Note 8.

17

Kitt Peak in the Quinlan Mountains as it appeared in 1957. Held of great significance in tribal lore by local inhabitants, the Papago Indians, the mountain was believed to be the home of I'itoi, the Papago Elder Brother. Today the site is dominated by more than a dozen highly sophisticated astronomical telescopes.

tered them. Ioligam figured prominently in tribal legend and was considered to be a favorite dwelling place of I'itoi (pronounced EE-E-TOY), the Papago Elder Brother. The mountain is still held of great significance in tribal lore.

The Papago people were less than thrilled about Kitt Peak being chosen as the site for a National Observatory. As Kitt Peak lay just inside the three million acre Papago Indian Reservation, permission was needed from the Tribal Council before work could begin. And this permission was not readily forthcoming. The Tribal Council was hesitant; they were unfamiliar with the science of astronomy, and were leery of an observatory being placed on their sacred mountain. That the presence of the observatory might despoil their land and upset their quiet way of life was among their concerns.

A cache of Papago pottery discovered on the summit in 1960 bears testimony to the significance of Kitt Peak in the Papago society. The cache contained the remains of miniature pottery vessels and a stone palette, believed to have been left as a shrine offering sometime between the years 1700 and 1860. Such shrines were rocky crevices in which offertory objects were deposited, but with the intent of restoring to Elder Brother (Nature personified) the balance which had been upset by the gathering of wild foods.

To illustrate, it is easy to visualize a Papago family on one of their periodic excursions to the foothills. After camp had been set up, the men would hunt and the women would gather saguaro, cholla, or perhaps basketry materials. The time would be a happy one. The people would be "living off the land" and the lack-luster staples of their diet (corn, squash, and beans) would be neglected in favor of the more "exotic", but seasonal, mountain foods. When the time had come for the return to the village, the people would remember the generosity of I'itoi, who created the wild things and who told the Papago where such foods could be found and how they might be prepared. They would then go up to the mountains where the gods live and leave a reciprocal offering as a token of gratitude for the gifts which Elder Brother had bestowed upon them.*

*See Reference Note 10.

19

Negotiations with the Tribal Council for permission to test Kitt Peak as a potential site for the National Observatory were lengthy. Permission was finally granted after Dr. Edwin F. Carpenter, director of Steward Observatory on the University of Arizona's Tucson campus, invited members of the Tribal Council and their families to look through the University's 36-inch telescope. The sight of the moon and planets awed the Council and convinced them that astronomical research would not despoil their land's natural beauty nor disturb their quiet peace. When Kitt Peak proved the optimal choice, the Tribal Council granted permission for the astronomers, whom the Papago call the **People with the Long Eyes**, to erect their telescopes on the mountain.

On March 5, 1958, members of the Schuk Toak District Council (on whose land the proposed observatory was to be built) met at the Santa Rosa Ranch and granted approval for the Papago Tribal Council to negotiate a long-term lease with the National Science Foundation.

A special bill was proposed allowing a perpetual lease for "as long as the land is used for astronomical study and research and related scientific purposes." The lease agreement was made binding in 1958 by a special law passed by the 85th Congress and signed by the President of the United States. The Papago received $25,000 for access rights to the site plus $10 an acre annually for about 200 acres of summit area to be then developed. The protective perimeter area of 2200 acres was to be rented at 25¢ an acre annually.

Site Development

...Jimmy Zumwalt was snaking a heavily laden water truck up the mountain... when the engine quit.... Jimmy jammed on the brake only to have the pedal snap off underfoot.... When he woke up he was sure he had made it to heaven. There he was lying on a grassy slope with a lovely waterfall playing nearby. With return to reality the grass became Spanish onion... and the cascade, his crumpled water truck....[11]

Dr. W.C. Livingston

The last hurdle had finally been crossed; now began the difficult task of converting a primitive mountain top into a first-rate modern research facility. Additionally, a modern facility with the necessary offices, workshops, and laboratories needed to support the observatory was to be constructed in Tucson. Tucson, once the Indian village of "Styuk-zon" is in the valley of the Santa Cruz River at an elevation of 2,450 feet. The Spaniards came to "Styuk-zon" before the Pilgrims landed at Plymouth, and in its long history the flags of Spain, Mexico, and the Confederacy have flown over it.*

---

*See Reference Note 8.

When planning of the observatory had been carried as far as was practicable by the panel appointed by the NSF, the Association of Universities for Research in Astronomy (AURA) was formed to carry the project foward on a permanent basis. Dr. Goldberg served as chairman of the organizing committee, and AURA was incorporated under the laws of the State of Arizona on October 28, 1957. The organizing universities were the Universities of California, Chicago, Michigan, Wisconsin, along with Ohio State University, Indiana University, and Harvard University; these seven universities are represented by the seven stars of the corporate seal.

The National Science Foundation approved a contract with AURA to carry out basic research in astronomy and to build, operate, and maintain the National Astronomical Observatory. Dr. McMath served as AURA's first president, while Dr. Meinel was appointed director of the NAO. The National Observatory dreamed of by Leo Goldberg half a decade earlier was becoming a reality.

The seven organizing universities of the Association of Universities for Research in Astronomy are represented by the seven stars in the corporate seal.

When the Papago Tribal Council gave permission for a "test site" on Kitt Peak in 1956, no road existed up the mountain. Pima County provided funds for a bulldozed trail to be cut from the Alambre Valley (near the base of the mountain) all the way to the summit. During the next 18 months, 30 tons of equipment and a small housing trailer were hauled up the primitive trail.

During the best weather the road was accessible to four-wheel drive vehicles, but during foul weather or for hauling equipment, only caterpillar vehicles could climb to the summit, owing to the high grades of up to 70%. The road was marginal even with respect to maintenance of the site survey tower and totally unsuitable for any construction phases of the project. Because wear and tear on the vehicles and personnel were considerable, and because excellent results of the seeing tests presaged much more future travel up the mountain, a new access road with no grades greater than 18% was needed.*

According to plans laid by the special panel appointed by the National Science Foundation, all phases of construction for the observatory were to be ready for initiation as soon as a final decision upon the site was made. Regardless of which site was ultimately selected, a paved public road was planned to extend to the observatory. But it was estimated to take eighteen months to budget and receive allocation of funds for the public road. Additionally, the regulations of the Bureau of Public Roads were so exact that the survey and engineering would require another eighteen months. Since it would take nearly three years after the final site was selected for construction of the public road, and since the site survey road was wholly unsuitable for construction, it was obvious that a construction access road was a necessity for the Kitt Peak site.[12]

Therefore, in 1957 a construction road was scraped and blasted to the summit by Tucson contractor Dick Kroecker. The route for the construction road was planned completely independent of the proposed route of the public road in order to avoid delays and inconvenience due to the latter's construction. A quonset hut was soon erected on the summit and further site testing commenced. Even if Kitt Peak had not been chosen as the

*See Reference Note 7.

Even the second road placed up the mountain — the construction access road — was so steep in spots it was necessary for heavily laden construction trucks to be pulled to the summit by a caterpillar tractor.

The quonset hut used to house and feed observatory personnel as it appeared in June, 1959. In 1961 the shelter was relocated to a meadow about one mile below the summit where it was converted into a storage building for seldom-used supplies and equipment.

26

site for the National Astronomical Observatory, the road and its high cost would not have been in vain, as the University of Arizona was seriously looking at Kitt Peak as a site for its own telescopes.

Because the truck access road had already been cut to the summit, when Kitt Peak was chosen as the site for the new National Observatory in March of 1958, construction of the observatory could begin immediately. That is, once lease negotiations with the Papago Tribal Council had been concluded. All construction materials for the #1 36-inch telescope, the 84-inch telescope, and the McMath Solar Telescope went up this road, some of the larger pieces caused considerable difficulty on the sharp turns. While the road was still steep and treacherous in spots, it was at least passable for nearly all vehicles.

Owing to the steep grades and the heavy volume of traffic (644 vehicle passages were recorded in March of 1960 alone), travel on the access road was hazardous. Eight miles of cable were laid

The sharp turns which existed on the construction access road caused considerable difficulty when hauling long loads up the mountain. A 30,000-gallon water storage tank is being cautiously hauled to the summit.

along the road from the summit office to a station at the base of the mountain. There a driver heading up the mountain could call the observatory to learn of any special road hazards which might exist, and also be informed of the presumed location of other traffic on the road. Despite the special precautions taken, numerous accidents still occurred. During October of 1959 two construction trucks turned over on the road; the following March a Travelall was severely damaged when it dropped some thirty feet after its driver missed a turn; and in November of 1960 no less than three accidents were reported. Fortunately, no one was seriously injured in any of the accidents, although many were considerably shaken up. One accident, ending with a rather humorous note, was recorded by Dr. Livingston, who was working at the observatory:

About a mile from the gate you will notice an ascending ridge which merges with the mountain just below the road. In 1961 Jimmy Zumwalt was snaking a heavily laden water truck along here when the engine quit just at this point. Jimmy jammed on the brake only to have the pedal snap off underfoot. Slipping backward he remembers futilely trying to wedge the accelerating truck against the rocky bank. When he woke up he was sure he had made it to heaven. There he was lying on a grassy slope with a lovely waterfall playing nearby. With return to reality the grass became Spanish onion . . . and the cascade — his crumpled water truck down there, fortunately, hung up on that ridge.*

Congress initially appropriated one million dollars for the new public access road in the 1958 session, but more money was needed. In July of 1959, Senator Carl Hayden reported the approval by the Appropriations Committee of both Houses of the necessary funds to complete the paved highway to the summit. Total appropriations amounted to $2.89 million, but would that be enough money to construct the highway? Fifteen bids were opened on November 24, 1959, for the total access highway project. The lowest bid came in at slightly over $1.8 million, while the highest bid was $3 million more.[13] However, the access road

---

*See Reference Note 11.

low bidder was released from the bid commitment by the Bureau of Public Roads following investigations of errors in the bid proposal. The second lowest bidder, Jones and Trotz of Albuquerque, New Mexico, with a bid of nearly $2.5 million was awarded the contract on December 24, 1959. Yes, there would be enough money to construct the new highway.[14]

Construction of the public road went smoothly and swiftly. On January 7, 1963, the new access highway was opened to the public. The 28-foot wide, 12-mile long highway with its smooth curves and gentle slopes is truly a joy to drive on. During 1963, the first year the observatory was open to the public, 58,884 visitors traveled up the road to see the telescopes.

Providing an adequate water supply for the observatory posed a challenge to the Kitt Peak engineers. Demand for water at the new facility would be considerable. Not only would the resident mountain staff and visiting astronomers consume precious water, but the darkrooms which were to be built on the mountain for processing photographic materials would guzzle water at an incredible rate. Additionally, hundreds of thousands of gallons of water would be used in the construction of the mountain facilities. Another factor concerning the water supply was the possibility of a major forest fire occurring on Kitt Peak. Evidence of past forest fires could be found on virtually all sides of the mountain. In order to combat a forest fire at the observatory an adequate reserve of water had to be stored on the summit, and a means of readily replenishing this reserve was mandatory.

The first thought regarding the water supply was to construct a dam in a natural drainage area which existed on the eastern slope about 1500 feet below the Kitt Peak summit. This idea proved impractical for a number of reasons including the low percentage of surface runoff and the high cost of constructing a suitable dam and reservoir. The necessary pipeline and pumping station to carry the water to the summit would also have been cost prohibitive.

A second, less expensive idea was to locate a spring near the summit and construct a pipeline with a pumping station. However, no suitable springs were found within 1500 feet elevation of the summit. Some evidence of seepage was found over the cliffs on the north side of the mountain, but the seepage occurred in an area which would have been extremely difficult to reach with a pipeline. There was also some doubt as to whether a spring could provide enough water for the thirsty observatory. Hence, this idea was also abandoned.

After reconsidering their approach to the water supply problem the engineers came up with a novel solution. They decided the best way to supply enough water for the observatory was to take the rain which fell on the top of the mountain and keep it there. Based on a yearly rainfall of only eight inches, the engineers' calculations showed that a paved collection area covering three acres would supply over half a million gallons of water yearly. Since the minimum rainfall recorded in the Kitt Peak area during the years 1930 to 1940 had been eight inches, with the average yearly rainfall more than twice that amount, such a collection area would indeed provide an adequate supply of water.

The primary catchment basin used to collect rainfall on Kitt Peak. The basin, adjacent to the visitors' parking lot, was covered with asphalt, then coated with four inches of concrete. The two water storage tanks, each capable of holding half a million gallons of processed water, are on the left, while the two buildings housing the purification system and pumping station lie at the lower left. *(Photograph by the author)*

Plans were completed and construction of the collection area was begun. A natural basin was selected on the summit, smoothed, and coated first with asphalt and then concrete. In addition to the paved catchment basin, a water purification and filtering system was built, along with the necessary pumping station. Garland Steel Company was selected to construct two storage tanks — each to hold half a million gallons of purified water. The two tanks, costing $45,000, were finished by the end of 1959.

As the observatory grew larger, in both facilities and personnel, there was some concern that the water collection area and storage tanks would not be able to meet the demand for water during extended periods of drought. To alleviate this potential problem, work commenced on the Horseshoe Valley Dam in May of 1963. A natural drainage area was selected below Horseshoe Ridge about one mile southwest of the summit. An earthen dam

Horseshoe Lake is a man-made reservoir which lies just one mile southwest of the Kitt Peak summit. Serving as a reserve water storage area for the observatory, the lake also supplies life-giving water to the mountain's wildlife. *(Photograph by the author taken at a point just below Horseshoe Ridge)*

31

was constructed and when the summer rains fell, Horseshoe Lake began to form. This lake, with a maximum depth of 24 feet and a capacity of six million gallons, provides a reserve water storage area near the summit which can easily be tapped during extended drought periods. Additionally, Horseshoe Lake serves members of the observatory staff who regularly visit the lake as a pleasant recreation area for picnics and fishing. Mountain wildlife also take advantage of this desert oasis, with tracks left by javelina, coati-mondi, deer, and mountain lion visible in the damp earth at the water's edge.

Early Telescopes and Facilities

The first major telescope to be erected on Kitt Peak was a 36-inch. Construction of the dome and building went smoothly with the exception of one minor problem, a local thunderstorm developed before the proper curing time had elapsed.

At the conclusion of the site survey, one of the site survey telescopes, the #1 16-inch, was overhauled and sent to Chile in April of 1961. There it took part in yet another site survey program which led to the establishment of the Cerro Tololo Inter-American Observatory. In March of 1959 the other site survey instrument was placed in the simple concrete block building shown in the photograph. This cylindrical structure was the first dome to be erected on the mountain, and within its walls the #2 16-inch site survey telescope was used for research purposes until September of 1962. The telescope was then moved to the Tucson headquarters where it participated in tests of possible drive systems for the 158-inch telescope. When the tests were completed, the 16-inch telescope was overhauled and then it, too, was sent to Chile. The dome for the #2 16-inch, now unoccupied, still stands atop Kitt Peak.

Two more 16-inch telescopes were destined for the observatory, both built by the Boller & Chivens Company of South Pasadena, California. The cylindrical housing for the #3 16-inch telescope was designed and constructed by Kitt Peak personnel, whereas the dome was built by the Ash Dome Corporation of Plainfield, Illinois. The base structure and dome were assembled in Tucson, transported to a site near the 84-inch telescope, and bolted to the prepared foundation. The #3 16-inch arrived during December of 1961, and by the following April the new telescope was fully operational.

A site near the #1 36-inch telescope was chosen as appropriate for the fourth 16-inch telescope to be built for the observatory. The building to house the telescope was designed by three members of the Kitt Peak staff: W. W. Baustian, Paul Loy, and

This cylindrical structure was the first dome to be erected on the mountain, and within its walls the #2 16-inch site survey telescope was used for research purposes until September 1962. The dome, now unoccupied, still stands atop Kitt Peak.

The cylindrical housing for the #3 16-inch telescope being bolted to a prepared foundation near the 84-inch telescope site.

The dome for the #3 16-inch telescope being lifted to the cylindrical housing during assembly at the site.

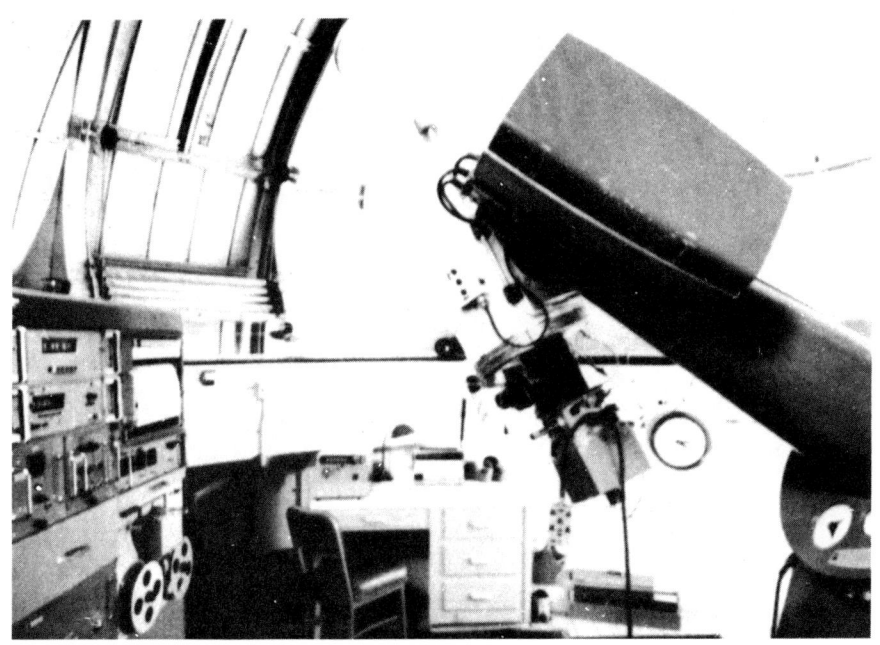

The #4 16-inch telescope is housed near the #1 36-inch telescope.

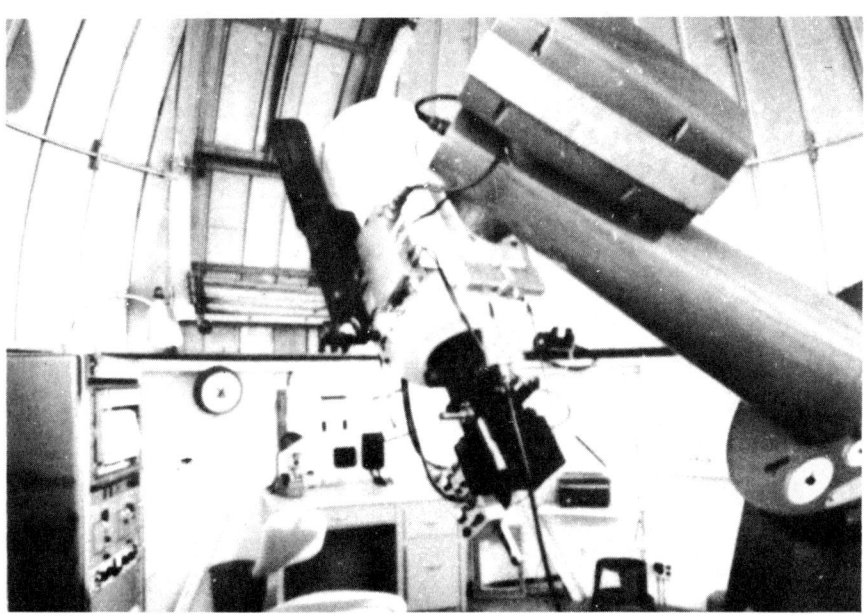

The #3 16-inch telescope.

Dave Crawford. The building and dome were quickly erected on the mountain by the M. Lang Construction Company of Tucson. The #4 16-inch telescope, visible in the photograph, was ordered without a tube and without optics. When the telescope mounting arrived, the tube was removed from the #3 16-inch and was placed on the #4 16-inch. In November of 1963 the #4 16-inch went into full-time operation.

A replacement tube for the #3 16-inch telescope was made by the Humbolt Instrument Company of Oakland, California, and was delivered in February of 1964. Optics originally intended for testing prototype instrumentation for the 84-inch telescope were added by observatory personnel and the #3 16-inch, shown in the photograph, went back into operation during March of 1964.[15]

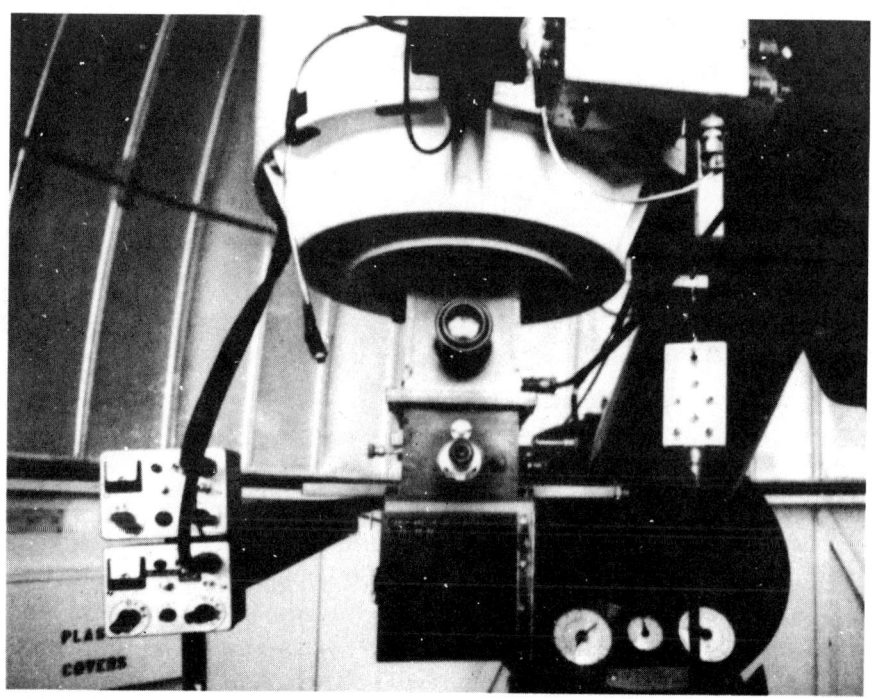

A photometer and early data acquisition system mounted on the #3 16-inch telescope.

Progress on the construction of the #1 36-inch dome in June, 1959.

The first major telescope to be erected on Kitt Peak was the #1 36-inch telescope. The mounting, built by the Boller & Chivens Company at a cost of $69,000, was delivered on February 26, 1960, and installation of the telescope was completed on March 12. The original mirror in the #1 36-inch was a ribbed aluminum blank plated with Kanigen. This mirror was replaced in 1963 and would later serve as the primary mirror for the #2 36-inch telescope.

View of the building and dome for the #1 36-inch telescope shortly after the $112,700 building was completed by the Murray J. Shiff Company.

Construction of the building and dome for the #1 36-inch telescope went smoothly with the exception of one minor problem. Approximately 80% of the exterior aluminum siding panels were damaged and had to be replaced. When the contractor applied the roofing material (consisting of a vinyl base acetate component, the function of which was to adhere together quartz crystals used in the roofing application) a local thunderstorm developed before the proper curing time elapsed. The rain and wind spread the material over the exterior siding, etching and scarring the surface.

During July and August of 1966 the telescope tube for the Cerro Tololo 36-inch telescope was interchanged with the #1 36-inch telescope. The old tube and optics went to Chile, and new optics made from Duran 50 were placed in the #1 36-inch telescope. In July of 1970 a new Cer-Vit primary mirror was placed in the #1 36-inch and the old Duran 50 optics replaced the metal mirror in the #2 36-inch telescope.

The #1 36-inch telescope. Note the skeletal structure of the 84-inch dome visible in the open slit of the dome.

Observers using the #1 36-inch telescope operate from a hydraulic floor which moves up and down to provide access to the telescope and instruments in all positions. This photograph shows the early data handling and instrument racks on the #1 36-inch observing floor. Since the racks moved up and down with the observing floor, occasionally a sleepy-eyed astronomer would accidently smash the telescope into the racks.

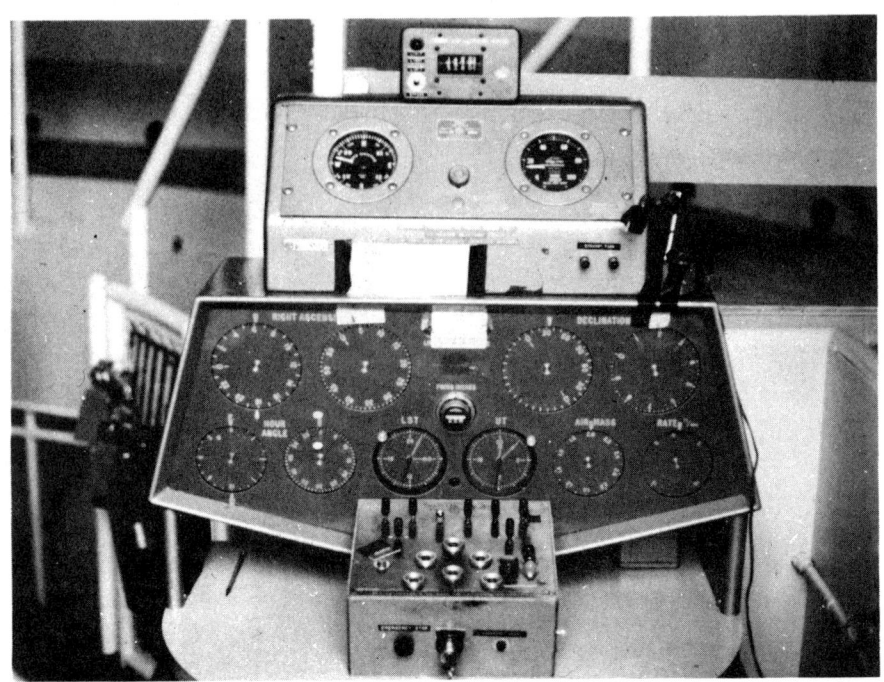

The control panel for the #1 36-inch telescope, unchanged since it was installed in 1960. While appearing quite complex, the panel is very simple to operate. *(Photograph by the author)*

The #2 36-inch telescope is housed in this dome on the mountain. *(Photograph by the author)*

42

Negotiations began with the Boller & Chivens Company in April of 1964 for two more 36-inch telescopes — one for Kitt Peak and one for the Cerro Tololo Inter-American Observatory in Chile. The Kitt Peak telescope arrived on June 27, 1966. By the end of 1966, the #2 36-inch telescope was in operation using the metal mirror originally installed in the #1 36-inch telescope.

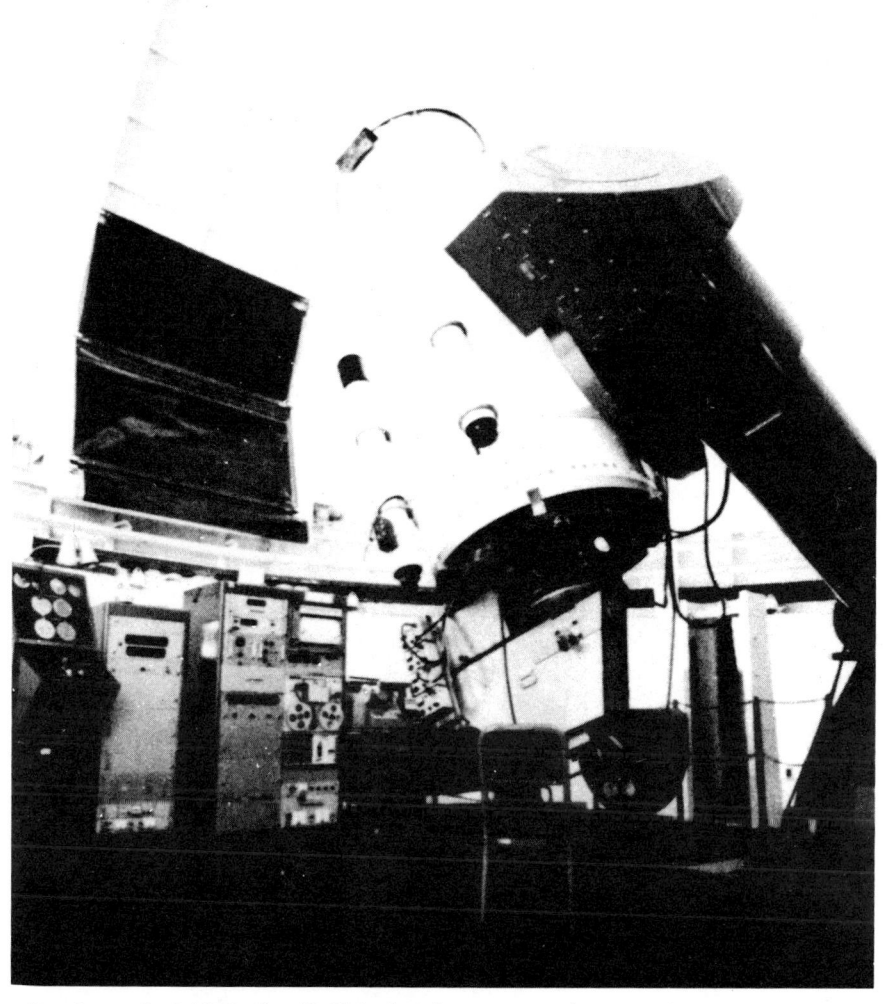

By the end of 1966, the #2 36-inch telescope was in operation using the metal mirror originally installed in the #1 36-inch telescope.

43

The 84-inch Telescope

*The next 15 months would see opticians
at Kitt Peak painstakingly grinding and
polishing the surface of the blank to an
accuracy of four-millionths of an inch.
But regardless of how careful one is,
accidents can occur....*

The second major telescope to be erected on Kitt Peak was, like the #1 36-inch, also designed for stellar research. While the specifications called for a mirror blank 84 inches in diameter, there was considerable doubt among the scientists as to whether the desired image size could be obtained using the full aperture. Hence for years, while being designed and built, the telescope was referred to as only an 80-inch. Not until the mirror was nearly finished and optical tests showed the entire mirror surface could indeed be used did the telescope receive the stature of being an 84-inch.

Corning Glass Works was awarded the contract for an 84-inch disk of pyrex glass late in 1958. A quarter of a century earlier Corning had pioneered the manufacture of giant-size mirrors when the company cast the 200-inch mirror blank for the Mount Palomar telescope. Physicist Dr. George V. McCauley, working at Corning, blazed the trail in large optics when he directed the complex job of melting, forming, and annealing the Mount Palomar disk. Another huge blank, 120 inches in diameter, was also cast as Corning tested its production methods in preparation for casting the 200-inch. The 120-inch mirror went into regular use at the Lick Observatory on Mount Hamilton in California.

This view from the southeast shows the 84-inch building as it appeared in April of 1960. On the left is the enclosure for the large coude spectrograph, to which was later added auxiliary optics making it functionally independent of the 84-inch telescope. The wing on the right served as a temporary museum for visitors from January of 1963, when the public road was opened, until June of 1964 when the museum was finished. Papago arts and crafts along with concessions were sold on the main floor which is now a laboratory for infrared detectors. Photographic displays were exhibited in the visitor's gallery near the 84-inch telescope, and descriptive photographs and models were displayed in what is now the Observer's office.

Unlike the 120-inch and 200-inch mirror blanks, which were made by ladling molten glass into molds, the 84-inch blank was to be cast by a relatively new procedure known as "sagging". In the new process large chunks of glass were to be placed in a mold and allowed to melt and fill the mold. Corning engineers claimed that "sagging" reduced the number of bubbles in the glass, was less complicated, and also less costly. Not only was the 84-inch disk to be the largest disk yet produced by the new method, it was also to be the first ribbed blank made by the new technique.

46

The Kitt Peak museum and visitor's center was finished in 1964. The front is decorated with a 5′ 8½″ high and 25′ 1″ long Mexican mosaic tile design of Mayan astronomical symbols and the famous Mayan observatory, the Caracol at Chichen Itza built in the 13th century A.D. *(Photograph by the author)*

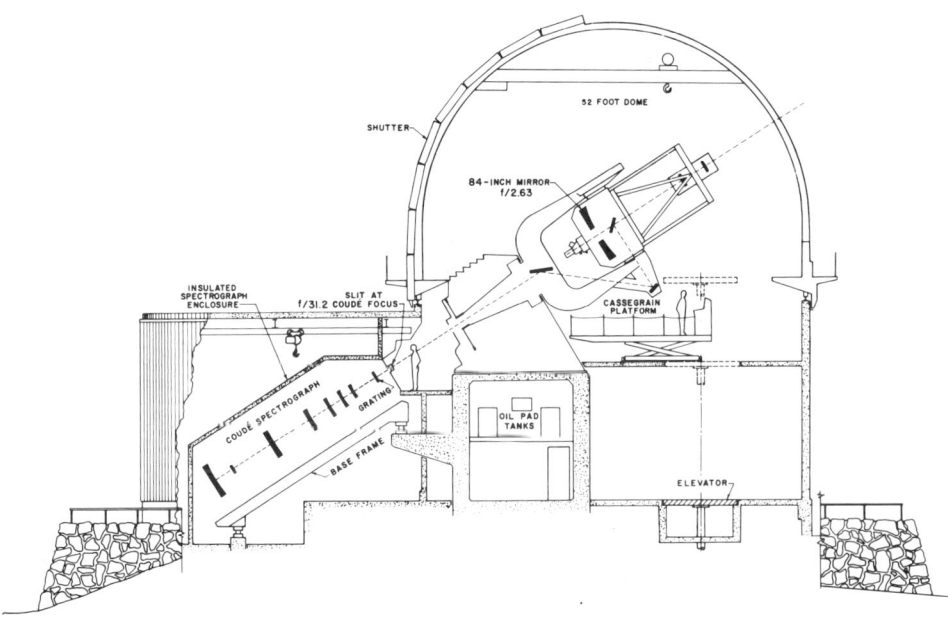

The optical configuration of the 84-inch telescope and its coude spectrograph. *(Courtesy of Kitt Peak National Observatory)*

To reduce the weight, yet retain the necessary strength and rigidity so important to astronomical mirrors, the disk was to be of ribbed construction and honeycombed on the back. Six months of planning and engineering were required before the disk could be cast. Although Dr. McCauley had retired in 1947 after a 30-year career in glass technology and research at Corning, he was called upon during all phases of the Kitt Peak 84-inch disk job.

One of the major tasks facing Dr. McCauley and the Corning crew was building the mold for the ribbed blank. To form the proper surfaces against which the mirror's flotation system would support the mirror in its cell, cores were constructed of ceramic brick. Using templates, each core was precisely placed in the mold, then cemented and bolted in position. A special cooling system was designed to prevent the bolts from melting in the intense heat while the disk was being formed.

The 2796-pound chunk of pyrex glass being carefully lowered onto the center core of the 84-inch mold. *(Courtesy of Corning Glass Works)*

When the mold was finished, nine chunks of glass were placed in the mold. The largest piece of glass weighed 2796 pounds and rested on the center core which was reinforced beneath by steel. Shortly after the furnace began heating up, this big chunk developed a crack. Since an early breakage of any sections of the glass might damage the delicate cores, the melting process was halted while the loosening pieces of glass were sliced off and placed in new positions around the mold.

The kiln was again heated up and this time the melting process proceeded satisfactorily. Under temperatures reaching approximately 2300° Fahrenheit, the separate pieces of glass melted to form the disk. An extra 48 hours of heating was used to eliminate slight bubbles, and the disk was then removed from the kiln and placed in an annealing oven. Controlled cooling was necessary to eliminate stress in the glass. The temperature of the glass was held constant for two months at 1040° F, then gradually the temperature was decreased and the disk was allowed to cool. The blank spent a total of seven months in the annealing oven.

Then began the ticklish job of removing the ceramic cores from the completed glass disk. The cores had to be carefully separated from the glass, an operation performed only after the mold bricks were removed and the disk was hoisted and turned over. A big question confronting the crew was whether the cores had adequately supported the huge chunks of glass and withstood the intense heat and fluid motions as the chunks had melted and filled the mold. If the cores had crumbled or were ripped loose, the glass blank could be ruined. Dr. McCauley's calculations said the cores would hold, and they did.

On October 1, 1959, the 2980-pound blank left Corning packaged in a felt-lined crate aboard a special freight car. New York Central Railroad carried the $115,000 disk to St. Louis; Southern Pacific Railroad then continued the haul to Tucson. The next 15 months would see opticians at Kitt Peak painstakingly grinding and polishing the surface of the blank to an accuracy of four-millionths of an inch.

Regardless of how careful one is during the grinding process, accidents can still occur. On June 12, 1961, during the last stages of fine grinding, opticians noticed a brief peculiarity with the grinding machine. They saw the 2100-pound grinding tool tend to stick and pivot in the central region of the 84-inch disk. Seconds later, one of the 260 pieces of tile which had been waxed to the face of the tool came loose and rode up on the surrounding beeswax. Thus, most of the weight of the grinding tool was then placed on this one tile, and since the machine was still in motion, the surface of the glass was gouged severely. A series of about twenty crescent-shaped cracks nearly 1/2 inch long and 1/8 inch deep were created perpendicular to the gouge. Additionally, when the loose tile overrode the rim of the inner hole in the blank, it chipped and broke down the edge of the hole.

The opticians believed that water trapped between the tool on top and an aluminum plate located one inch below the surface of the central hole had caused the accident. The water had appar-

The "sagging" process used by Corning to cast the 84-inch disk resulted in a blank of uniform hardness which permitted opticians to grind, polish, and figure the mirror in only 15 months, compared with the normal four years required with conventional disks.

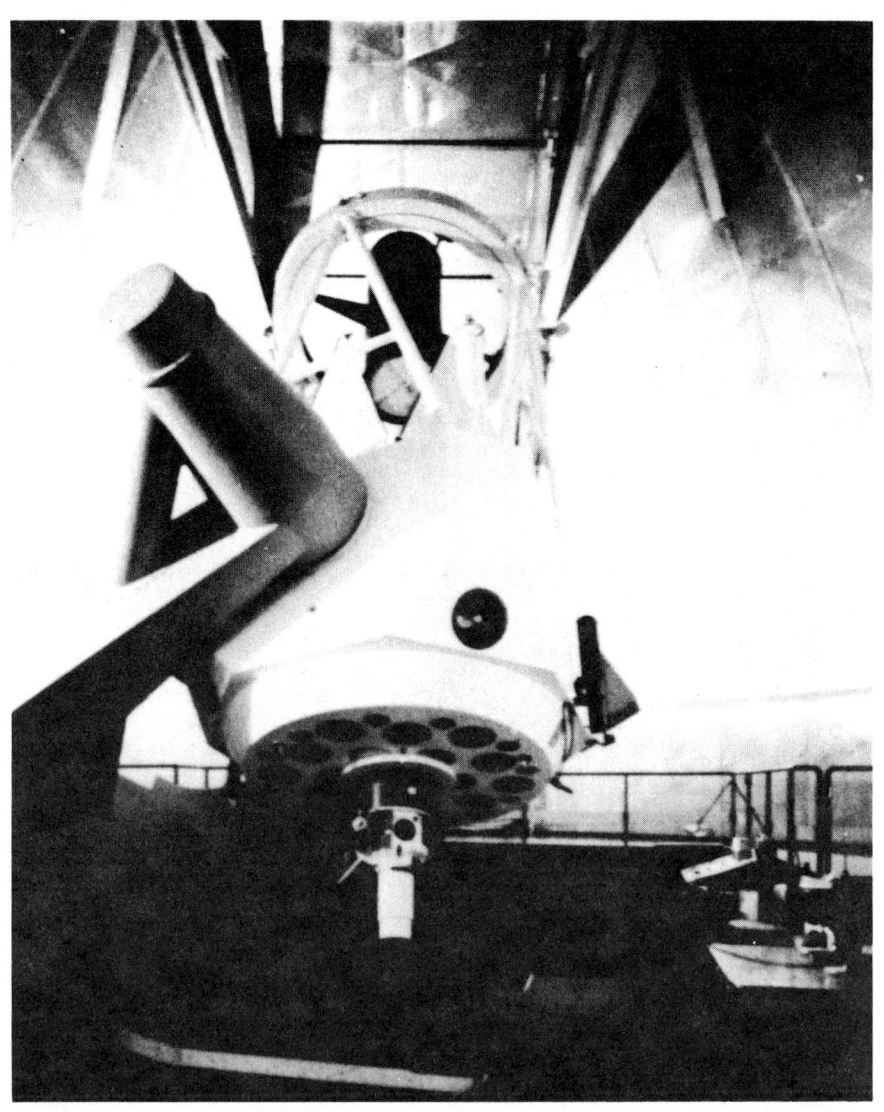

The 84-inch telescope with fork mount and hydraulic floor.

ently flushed the surface of the central region free of grinding abrasive and caused the seizing. To avoid a similar occurrence the Optical Shop eliminated the aluminum plate which had been installed to prevent abrasive and water from getting down into the rubber support pads of the grinding machine. Additionally, the central hole in the 84-inch mirror blank was enlarged from 24 to 24 1/2 inches, and opticians reground the disk to remove the scratches.[16]

Meanwhile, work was progressing rapidly on the mechanical parts of the telescope, designed by Aden Meinel and Bill Baustian. Willamette Iron and Steel Company, based in Portland, Oregon, received the $363,313 contract in July of 1959. The 70-ton reflector required two and a half years of engineering and

These two photographs taken in November of 1961 show the polar axle base frame and the fork assembly for the 84-inch telescope being lifted up through the open shutter of the dome. The mechanical parts of the telescope were carefully positioned on the observing floor within the dome and the telescope was reassembled.

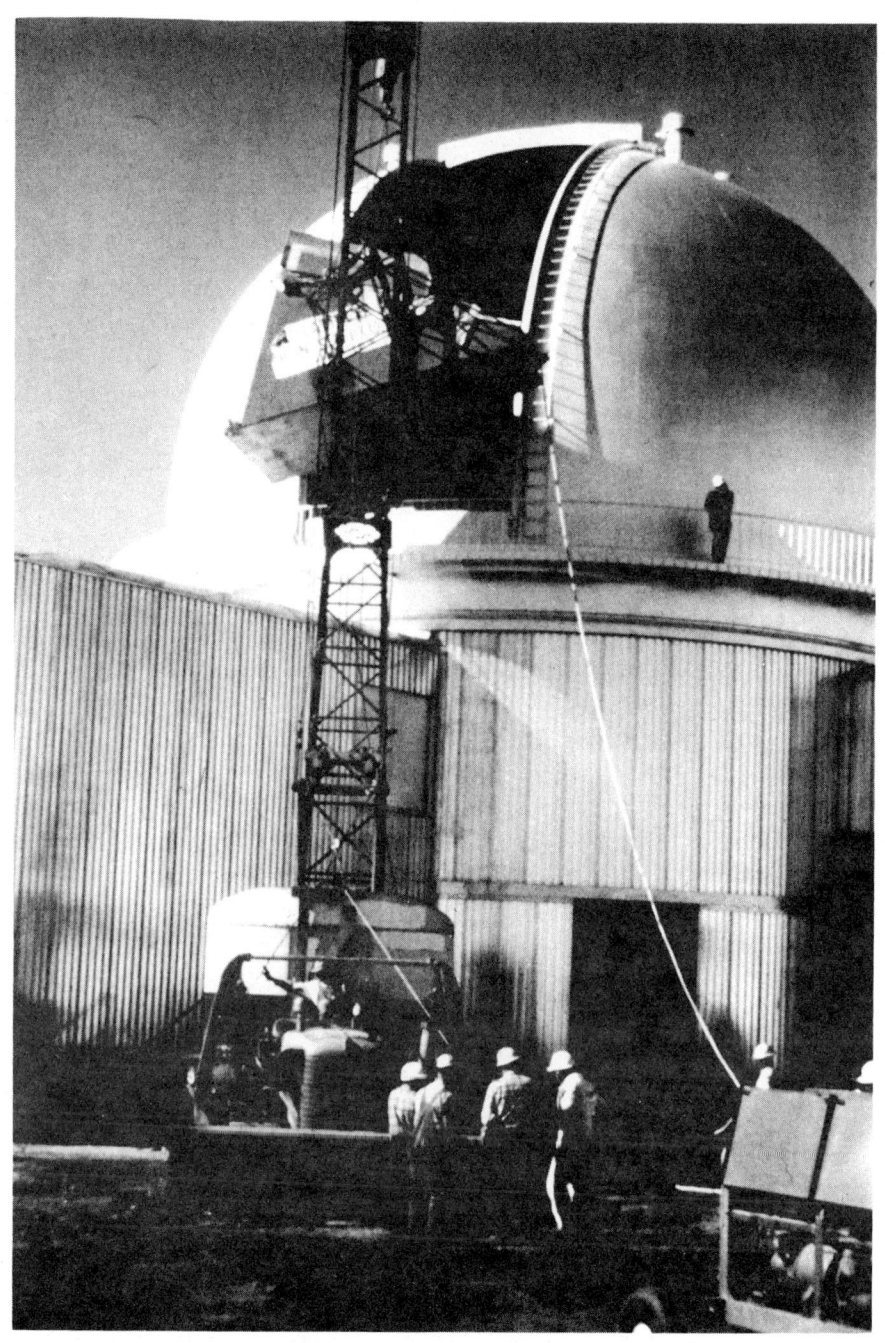

Workers installing the polar axle of the 84-inch telescope. The north or upper support of the polar axis is large enough for oil-pad bearings, whereas the south polar support is small and requires only standard bearings. A single worm wheel 90 inches in diameter with 720 teeth is used both for slewing and tracking in right ascension.[17]

This view from the east, shows the housing for the 84-inch telescope and coude spectrograph, with the dome for the #3 16-inch telescope visible on the left. In November, 1969, the Coude Feed Auxiliary project was begun. A 60-inch flat, formerly used as a heliostat mirror in the McMath Solar Telescope, is used in a computer-controlled mount located above the spectrograph housing. Light is directed from this flat to a 36-inch diameter image-forming mirror located in the top of the cylindrical tower seen at the right of the #3 16-inch dome. The 36-inch mirror sends the light to a small third mirror centered on the polar axle of the 84-inch telescope which then directs the light through the spectrograph slit to the collimator. *(Photograph by the author, 1977)*

shop work. Late in October of 1961 the mechanical parts were shipped from Willamette to Kitt Peak. The individual components of the telescope tube, fork, and base assemblies were hoisted through the open slit of the dome and carefully placed on the observing floor. During December the telescope was reassembled in its new home.

The 84-inch telescope has a fork mount which requires a hydraulic floor that can be moved sideways as well as up and down, in order to clear the fork when the telescope is pointed at large angles. The length of the fork arms depended in part on the amount of clearance required for attaching instruments. The design of the 84-inch telescope permitted the primary mirror to be very close to the declination axis, so an inside fork length of only 9 feet gives more than 4 feet of clearance. This design requires an internal truss to allow the mirror cell to deflect under gravity by the same amount as the secondary support ring. While the center section box frame seems to hold the mirror cell, the latter is actually attached to a parallelogram truss hidden inside and fastened to the upper flange of the box frame, near the junction of the truss from the upper ring holding the secondary mirror.*

Construction of the 84-inch dome by the Murray J. Shiff Construction Company of Tucson had progressed rapidly and with only one slight problem. In November of 1959, during a routine examination, the North-South center line of the telescope pier was discovered to be off about 1.5° NNW of where it should have been. However, this error was not critical since the sole plate of the telescope would rotate about its true center point to the desired angle.

The 84-inch telescope was officially available for use by visiting astronomers on September 15, 1964. The first regularly scheduled visiting astronomer was Dr. G. Van Biesbroeck, from the Lunar and Planetary Laboratory at the University of Arizona. Dr. Van Biesbroeck used the telescope visually for micrometer measurements of close double stars. The telescope lived up to the expectations of the Kitt Peak staff, and remains one of the most heavily demanded telescopes on the mountain.

---

*See Reference Note 17.

The 84-inch dome. The tall tower to the right of the photograph houses the image-forming mirror for the coude spectrograph. Behind and partially obscured by the tower is the dome for the #3 16-inch telescope. The McMath Solar Telescope is visible to the left of the 84-inch dome. *(Photograph by the author, 1979)*

Star trails behind the 84-inch telescope. *(Photograph by the author)*

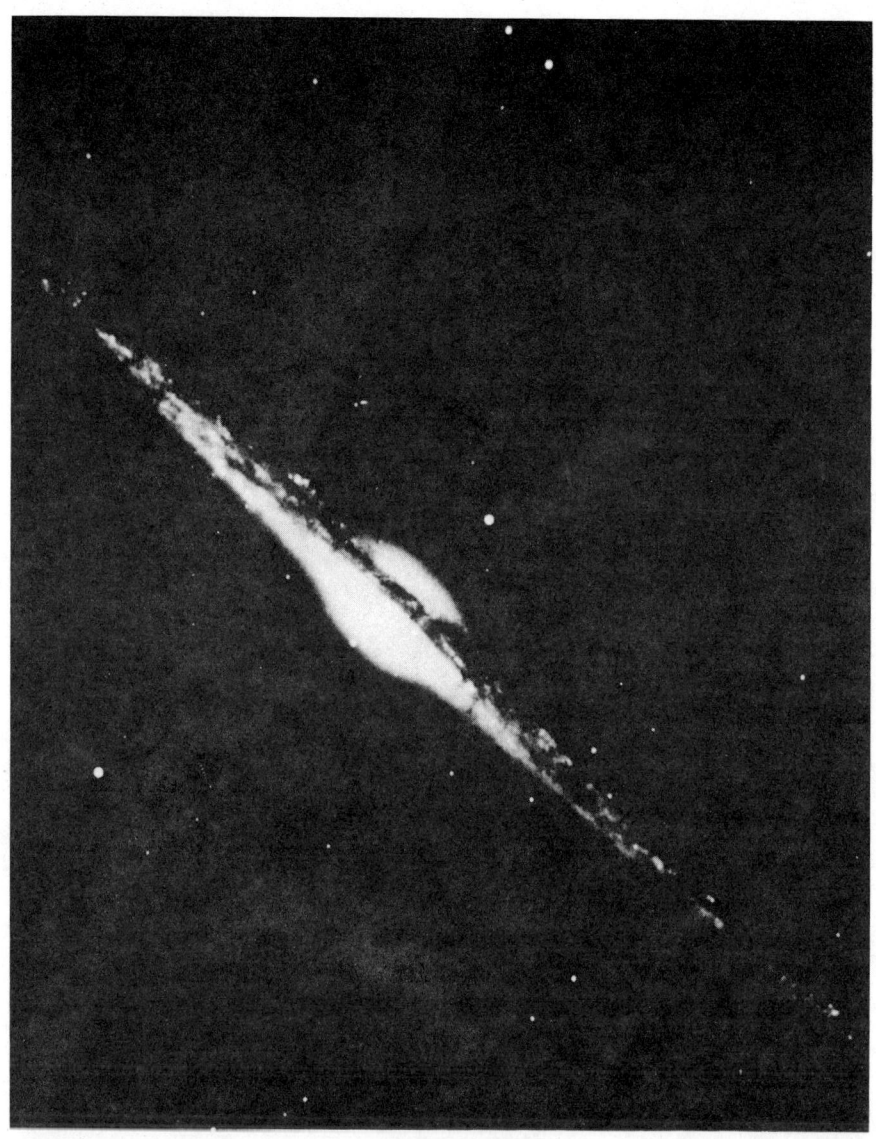

This spectacular galaxy, NGC 4565 in the constellation Coma Berenices, is one of the finest examples of an edge-on spiral galaxy visible to our eyes. This galaxy, home to billions of flaming suns, is receding from Earth at 750 miles per second. The distance between NGC 4565 and the earth is so great that starlight leaving the galaxy must travel for nearly 20 million years before reaching our eyes. Hence, when we view the galaxy we are not seeing it as it looks today, but as it appeared nearly 20 million years ago. This photograph was taken with the 84-inch telescope.

The McMath Solar Telescope

The great new solar telescope at the Kitt Peak National Observatory in Arizona is a source of pride to the nation....

Bold in concept and magnificent in execution, the instrument is the crowning achievement of the career of the late Robert R. McMath....

The thanks of the nation go also to the Papago Indian Tribal Council, and members of the tribe, for welcoming the National Observatory to Kitt Peak. This mountain, notable in the ancient lore of the tribe, will now have a salient role in the modern exploration of the sun and the universe of stars.

John F. Kennedy

When Kitt Peak National Observatory was dedicated on March 15, 1960, the ground-breaking ceremony took place for what would become the world's largest and most sophisticated telescope for studying the sun. The telescope, monstrous in proportions, would incorporate a tower rising nearly one hundred feet above the ground from which a shaft would slant two hundred feet to the ground where a tunnel would continue an additional three hundred feet into the mountain. Before construction of the solar telescope was complete, the National Science Foun-

◄— Steel reinforcement being placed for the solar telescope tunnel. The concrete pier to support the heliostat has already been poured. *(Photograph by J. C. Golson)*

dation would spend nearly four million dollars on the project, and tragically, one man would lose his life in a construction accident at the site.

The optical specifications designed by AURA personnel called for a heliostat mirror measuring 80 inches in diameter. To improve the "seeing", this mirror was to be supported one hundred feet above the ground. Sunlight would reflect off the heliostat mirror to an image-forming mirror 60 inches in diameter located nearly five hundred feet away. The light would then reflect back to a 48-inch diameter flat mirror which would bounce the light to underground observing rooms. Astronomers would be able to study an image of the sun nearly a yard across or, by means of auxiliary optics, produce a spectrum 70 feet long.

Plans for this mammoth telescope had begun years earlier when the special advisory panel appointed by the National Science Foundation had advocated construction of a solar telescope in 1955. During the fall of 1957 Detroit engineer W. Zabiskie prepared detailed drawings of three possible structures. In all three of these designs the telescope was an immense vertical triangle with its hypotenuse along the polar axis. The telescope therefore appeared much like the gnomon of an incredibly large sundial. A year later, when further design work was started, the Chicago firm of Skidmore, Owings, and Merrill was asked to study all possible structures, including those already designed by Zabiskie, that would satisfy the particular optical requirements for the solar telescope.[18]

Designing the solar telescope was a difficult task. The structure supporting the heliostat mirror had to be so rigid that even when a 25-mile-an-hour wind slammed against it, the image of the sun at the end of the 780-foot optical path would not deflect by more than 1/60 of an inch. Additionally, to avoid thermal effects on the optical path, the air inside the structure would have to be maintained at a temperature equal to the air outside. Therefore, a design criterion was that all surfaces exposed to sunlight had to be temperature-controlled.

Skidmore, Owings, and Merrill prepared ten possible structures and from these recommended two potential designs. One design consisted of a tapered cantilever tower projecting from the ground toward the star Polaris with a similar inner section supporting the heliostat mirror, appearing much like an extreme case of the leaning tower of Pisa. The other proposed design favored a vertical concrete tower supporting the heliostat with surrounding wind shield and a diagonal wind shield for the optical path. The vertical tower design was less costly and more stable than the cantilever design, and hence was adopted when AURA representatives met with the Chicago firm in June of 1959.

In the chosen design, the heliostat would sit atop a massive concrete cylinder 26 feet in diameter with steel-reinforced walls four feet thick. Surrounding this pier would be a water-cooled structure to deflect the wind and maintain temperature stability. Attached to this wind shield was a diagonal structure, again water-cooled, encasing the light path from a point just below the heliostat mirror all the way down to the ground.

Projecting from the ground on the eastern edge of the Kitt Peak summit area, the McMath Solar Telescope resembles the gnomon of a huge sundial. *(Photograph by the author)*

Western-Knapp Engineering Company, a subsidiary of Western Machinery, was the prime contractor for the solar telescope. Excavation of the 300-foot tunnel and adit was perhaps the most dangerous task in constructing the mammoth telescope. All the necessary precautions were taken to ensure the construction workers' safety. As the tunnel was extended deeper into the mountain, rock bolts were driven into the sides of the tunnel and steel netting suspended to catch any rocks which might fall. But, man can not always protect himself from his machines, and a tragic accident occurred when an end-loader (used to remove boulders from the tunnel) rolled over and instantly crushed the driver to death.

Excavation work for the McMath Solar Telescope. The top ten feet of the rock was quite decomposed and was easily removed using a caterpillar tractor with a ripper blade.

The skeletal framework of the diagonal wind shield is visible as construction progresses on the solar telescope. Installation of the tube-in-strip cooling panels had already begun on the vertical wind shield when this photograph was taken. Nearly 30,000 square feet of the cooling panels were installed over the framework of the telescope. Through the panels nearly 19,000 gallons of a special antifreeze solution are circulated.

A major engineering problem to be overcome was designing a system for adequately cooling the massive structure. Experiments conducted by Mr. and Mrs. Raymond Bliss of the University of Arizona's Solar Energy Laboratory shed some light on the problem. The cooling load for the solar telescope's outer skin, with its 34,000 square feet of area exposed to the sun, was tremendous. But, if the skin were painted white with titanium dioxide paint, the Bliss's found, only 14 tons of refrigerated solution would be required.*

Six brands of titanium dioxide paint were chosen and tests were conducted to determine their suitability. Each type of paint was applied to two 12-inch disks of aluminum and the amount of solar energy absorbed by each was measured. One disk of each paint was placed on Kitt Peak for one year of weathering after which the relative absorption of the paints was redetermined, and the final paint selected.

---

*See Reference Note 18.

One of the most complex and difficult construction problems was fabrication and installation of the tube-in-strip cooling panels. Nearly 140 of these 34-foot by 8-foot copper panels comprise the outer skin of the telescope. Through these panels are circulated an impressive 3,600 gallons of antifreeze and an incredible 15,000 gallons of water. A refrigerating plant located about two hundred feet away pumps the chilled mixture to the telescope through underground pipes.

However, the tube-in-strip panels provided only part of the answer. During the summer when the tunnel (ground temperature 55°F) was cooler than the ambient air, there would be no convection in the telescope. However, during the winter when the tunnel was warmer than the ambient air, a stable column of air in the telescope could only be obtained by incorporating a refrigerated liner in the tunnel portion below ground. Therefore, nearly twenty-five thousand feet of one-inch diameter pipe on six-inch centers was installed in the tunnel portion. A three-inch blanket of insulation was placed behind the pipes, and the front was covered with aluminum sheets.

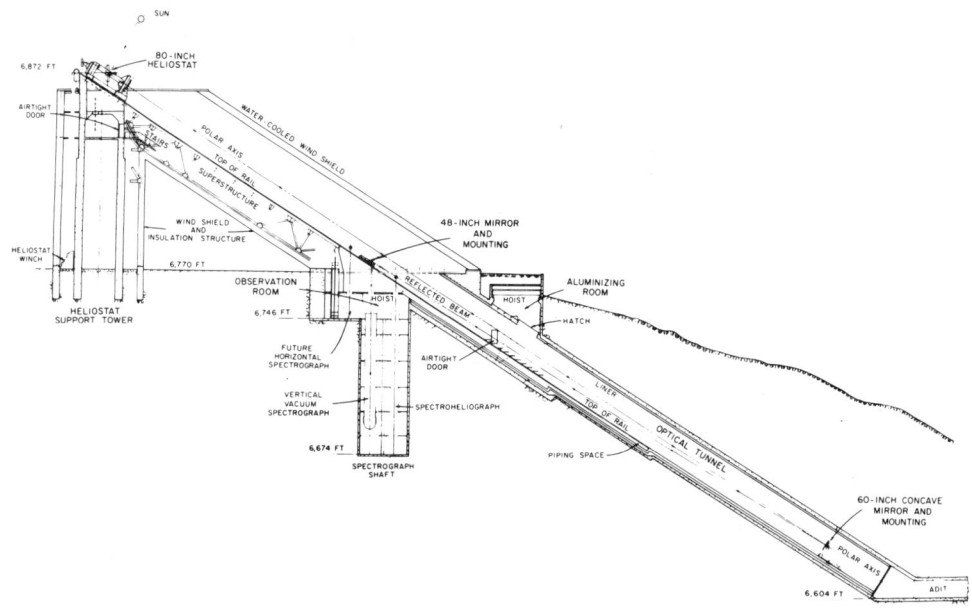

Outline drawing showing the optical path in the McMath Solar Telescope. *(Courtesy of Kitt Peak National Observatory)*

Construction workers were preparing the foundation forms for the southern solar observing room when this photograph was taken. The three large ports direct light from the solar telescope to various instruments in the observing room.

The heliostat base frame tips the scales at an enormous 27 tons, and had to be manufactured in four parts. The frame carries four strong unsprung wheels and three jacks for polar alignment. The frame, as well as the other mirror mountings, rests on a 12-gauge steel track which extends nearly one-tenth of a mile from the heliostat to the primary mirror. This track provided a convenient means of handling each mirror mounting during the initial assembly and also whenever the mirrors are to be washed or aluminized. Hoists can move the heliostat along the track at four feet per minute.[19]

The original optics in the solar telescope included two quartz disks with an interesting history. These blanks were the final efforts of A. E. Ellis and E. Thomson, at the Lynn plant of General Electric in 1931, to prepare quartz disks up to 200 inches in diameter for the Mount Palomar telescope.*

*See Reference Note 18.

65

At that time the 60-inch disks were considered useless for astronomical purposes. For many years after the work ended, they remained at the General Electric plant. Nobody wanted them, nobody cared. Recently, however, Robert McMath of the McMath-Hulbert Observatory at Pontiac, Michigan, who has done remarkable work making motion pictures of the sun, remembered the disks. Perhaps, as Hale had suggested, they might be usable for solar work .... Today the disks are in Ann Arbor. Perhaps, after all, they may be useful, and those years of "arduous" work will not be wasted. If so, in the light of their original fantastic cost, they will be the most valuable mirrors in use anywhere in the astronomical world. [20]

For many years the two glass blanks lay undisturbed in the yard at the Lynn plant. Around 1950 General Electric notified Dr. Bowen at Mount Palomar that the company wished to dispose of the disks. Aware of Dr. McMath's dream of a large solar telescope, Bowen offered to turn the disks over to him for the cost of moving them from Lynn to Ann Arbor. Dr. McMath sent a truck to transport the disks to the University of Michigan Observatory, where they remained stored until selected as mirrors for the new solar telescope.

One of these disks was the first 63-inch diameter quartz disk to be cast by the General Electric Company. After being sprayed with a one-inch layer of molten quartz the disk spent nine days in the annealing oven — or almost nine. "Someone, somehow — nobody quite knew who or how — some member of the big staff opened the oven just a few hours ahead of the deadline .... The disk was cracked!"[21] Ellis set himself to the task of repairing the crack in the disk.

With glass he never would have attempted it — patching up a crack in a five-foot slab. It would have smashed the minute the heat was turned on. But with quartz it ought to be easy. With a blow-torch and care new material might be melted into the break and welded in place, just as if it had been steel .... Everything went well. Up rose the temperature in the electric oven till the disk was hot. With a special torch he had devised for the purpose Ellis began filling in the crack in the quartz. The disk accepted the new material without a murmur. Ellis worked all day and into the night. He was nearing the end and a great load was slipping off his shoulders. If he could indeed weld up a cracked disk and make it like new, there would be one more thing that quartz could do and glass could not. Then at midnight the electric current failed. Somebody had blown out a switch in the power house. The whole plant was dark. The heating units in the oven went out and the disk began to cool rapidly, without control. The quartz stopped welding. The crack, so nearly closed, split open again, widened, grew serious, sprang with a dull thud across the whole expanse of glass .... The crack was not serious now, it was fatal.[22]

The McMath Solar Telescope is actually three telescopes in one. In addition to the main heliostat, two smaller mirrors, referred to as the East and West Auxiliaries, can independently collect and deliver light to the various instruments in the observing rooms. The 82-inch heliostat mirror is shown open for observing while both the East Auxiliary and West Auxiliary remain closed. The back of the heliostat mirror is supported by air pressure, less than one pound per square inch being required when the mirror is horizontal. The added pressure is decreased as the mirror is tipped and must be replaced with a slight vacuum when the mirror faces downward. *(Photograph by the author)*

The disk was eventually successfully welded back together and, after being reduced in size to a diameter of only 48 inches, served as the #3 mirror in the optical system of the solar telescope. The mirror began to show a slight warping along the weld, and was replaced in 1965 by a 48-inch diameter flat beryllium mirror.

The second fused quartz disk cast at the General Electric plant served as the main heliostat mirror in the solar telescope until 1966 when it was replaced with a larger 82-inch diameter quartz mirror. The 63-inch diameter mirror was not to be retired, however. When the Coude Feed Auxiliary at the 84-inch telescope was erected, the mirror was placed as the light-gathering flat in the optical system.

On November 2, 1962, the solar telescope was formally dedicated. Speakers at the dedication ceremony included Dr. Alan T. Waterman (Director of the National Science Foundation), Dr.

Looking down the dizzying depths of the solar telescope one can see the 60-inch image forming mirror of the main heliostat, and the two smaller image-forming mirrors of the Auxiliaries. Behind the aluminum panels which line the tunnel lie nearly 25,000 feet of tubing through which a refrigerated solution is circulated. *(Photograph by the author)*

Nicholas U. Mayall (Director of Kitt Peak National Observatory), and Mr. Enos Francisco (Chairman of the Papago Tribal Council). The telescope was officially named for Dr. Robert R. McMath, who had passed away on January 2, 1962, just ten months before the telescope was dedicated.

Dr. McMath, professor emeritus of astronomy at the University of Michigan and director of the McMath-Hulbert Observatory, had served as chairman of the special NSF advisory panel for the National Observatory. His special talents in organizational matters and his acumen in instrumentation helped materially to speed up the establishment of Kitt Peak National Observatory.[23]

The following letter from President Kennedy was read at the dedication ceremony by Dr. Waterman. The original letter hangs on a wall in the solar lobby.

The great new solar telescope at the Kitt Peak National Observatory in Arizona is a source of pride to the nation. The largest instrument for solar research in the world, it presents American astronomers with a unique tool for investigating the nearest of the stars, our sun. This project is of exceptional interest to all our citizens, for the observatory is financed by the Federal Government through the National Science Foundation, and is available to qualified scientists with meritorious programs of research.

Bold in concept and magnificent in execution, the instrument is the crowning achievement of the career of the late Robert R. McMath, builder of solar telescopes, for whom it is named. I extend the gratitude of the nation to Dr. McMath's family and especially to his wife, Mary Ridgely McMath.

The thanks of the nation go also to the Papago Indian Tribal Council, and members of the tribe, for welcoming the National Observatory to Kitt Peak. This mountain, notable in the ancient lore of the tribe, will now have a salient role in the modern exploration of the sun and the universe of stars.

John F. Kennedy

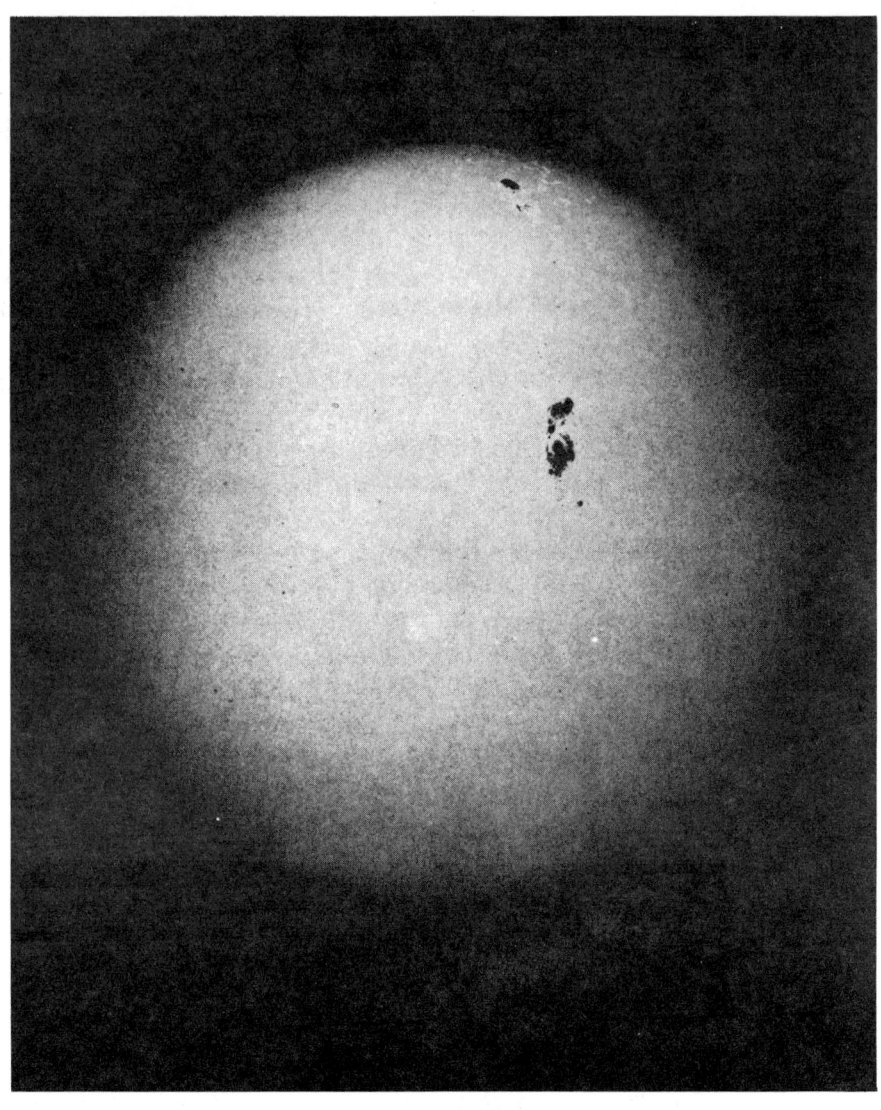

Photograph of the sun taken with the McMath Solar Telescope on July 4, 1974. The dark areas, called sunspots, on the visible surface of the sun are temporary cooler regions associated with magnetic disturbances.

Constructed in 1973, the Solar Vacuum Telescope was built in response to the need of the solar physics community for synoptic maps of magnetic activity. The telescope was completed in time to provide valuable support for the Skylab Mission. Although the sun is 93 million miles from the earth, these two solar telescopes are capable of distinguishing features on the sun's surface a mere 400 miles across. The small dome at the right edge of the photograph houses a small telescope which studies the sun in hydrogen-alpha light.

View toward the east. Appearing behind and to the right of the solar telescope is the 75-foot-high rectangular concrete tower of the Solar Vacuum Telescope. *(Photograph by the author, 1980)*

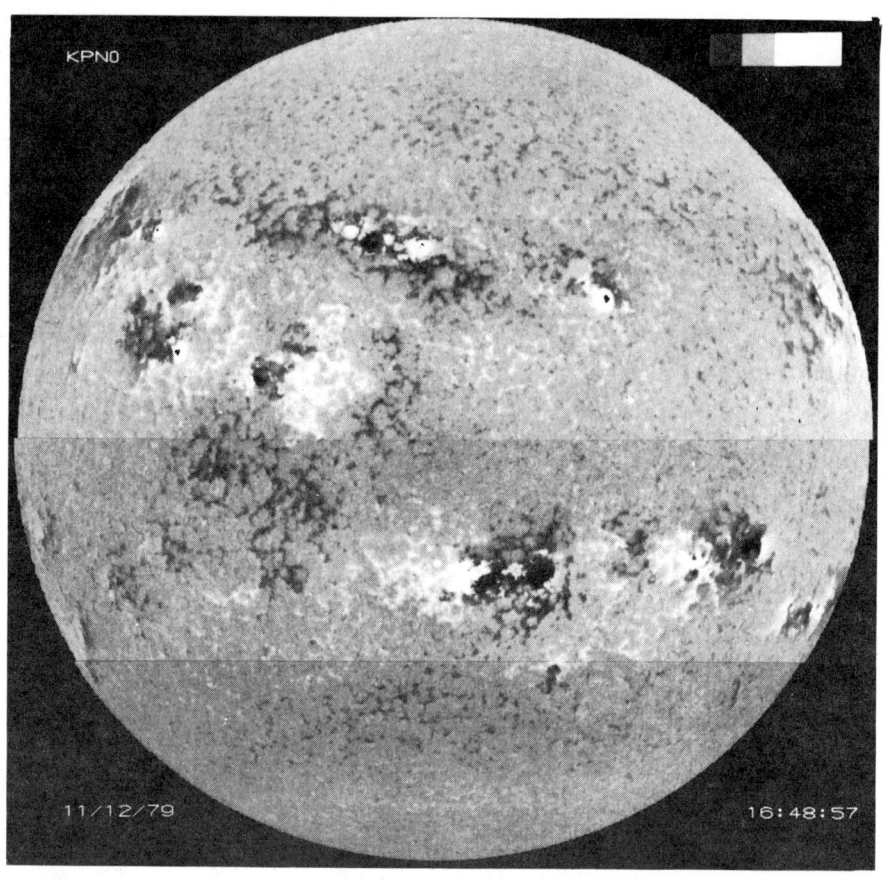

This solar magnetogram was taken in November of 1979. Astronomers at Kitt Peak study daily the ever-changing magnetic fields on the sun, alerting colleagues at other observatories of any unusual activity.

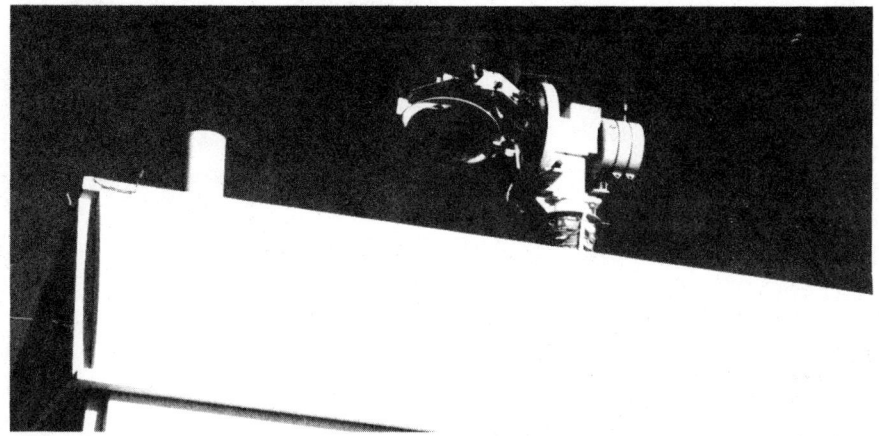

One of the mirrors used to direct sunlight down the Solar Vacuum Telescope. A 41-inch diameter coelostat mirror directs the light to this 36-inch flat mirror which in turn sends the light vertically downward into a large vacuum tank inside the tower. A 24-inch image-forming mirror and most of the light path are inside the vacuum tank, thus improving image stability. *(Photograph by the author)*

The lights of Tucson appear on the horizon. *(Photograph by the author)*

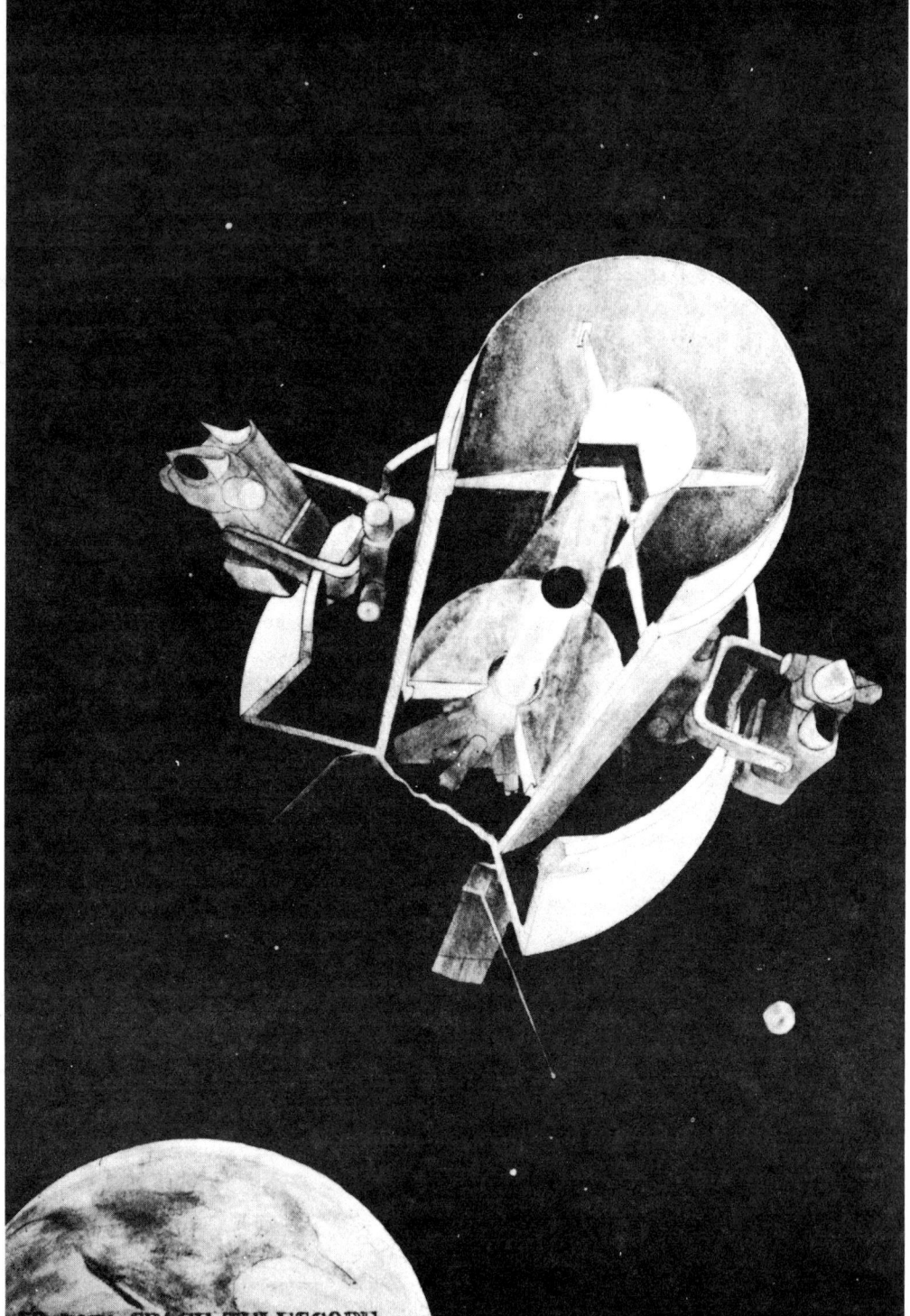

50-INCH SPACE TELESCOPE
1959   A.B. Meinel

Kitt Peak and the United States Space Program

The Kitt Peak Space Division assisted the United States in its space program including early studies on a space telescope and the White Sands Aerobee launch program.

In addition to designing and building new telescopes and instrumentation for ground-based astronomy, for more than a decade Kitt Peak National Observatory was actively involved in the space program. Kitt Peak scientists not only designed instruments to conduct frontier astronomical research above Earth's obscuring blanket of air, clouds, and dust, they assisted in the development of the United States space effort. The efforts of the Kitt Peak scientists materially assisted the United States in achieving supremacy in space flight.

The Kitt Peak Space Division became active on May 27, 1959, when the National Science Board approved an AURA budget request in the amount of $160,000 to do preliminary design work for a large aperture orbital space telescope. More momentum came in 1960 when the National Science Foundation granted an additional $252,000 to the observatory for space research. Over the next decade, hundreds of thousands of dollars were to pour into the department before funds suddenly dried up.

◄— An early conception of the 50-inch Space Telescope, prepared by Dr. Meinel in 1959. Solar cells were to provide power for such instruments as spectrometers and photometers.

The Space Division was concerned primarily with two major programs, one of which was the Space Telescope Study. Although not aimed at the immediate construction of a space telescope, the study did represent an effort toward a later generation of orbiting telescopes. As early as 1959 the orbiting telescope was envisioned as having a mirror 50 inches across, and accommodating five separate experiments: ultraviolet and infrared spectrometers, an extended range multicolor photometer, a bolometric photometer, and a variable magnification television system for high-resolution spatial studies of celestial objects.

Solar cells were to serve as the power source for the satellite, with the cells distributed on the rear face of the telescope platform. The Space Telescope would spend about eight hours out of every 24-hour period with its rear face pointed to the sun for recharging batteries. Designers anticipated that solar research of a simple nature could be conducted while the batteries were recharging.

While the Goddard Space Flight Center was more intimately associated with the design of such an orbital telescope, Kitt Peak was encouraged to construct prototype mirrors for the telescope. Another necessary phase in the development of the Space Telescope was the successful operation of an "earth" telescope from a remote control point, and it was in this capacity that the Kitt Peak Space Division played a very critical role.

While this project was originally concerned only with engineering research directed toward eventual space-flight applications, the project soon emphasized the development of a fully automatic, remotely controlled telescope for ground-based astronomy. What was envisioned was a remotely controlled telescope system capable of automatically executing a program of astronomical observations and performing limited data reduction. Although the Remote Control Telescope (RCT) was initially planned with a mirror 36 inches in diameter, the size grew to 50 inches when a 50-inch metal mirror blank was acquired.[24]

The Boller & Chivens Company received a $69,250 contract for the RCT mounting in 1961. The mounting was essentially like the one designed for the #1 36-inch telescope, but with additions for digital readout of the telescope positions and digital telescope drives. The mounting was delivered on October 22, 1962, and installed in temporary quarters adjacent to the warehouse at the Tucson offices, where it underwent extensive testing. Until the 50-inch optics were finished, the mounting was temporarily fitted with a 16-inch optical system from one of the site-survey telescopes. A photelectric finder and automatic UBV photometer were installed to facilitate testing the mounting.

The 50-inch building was constructed by the Murray J. Shiff Construction Company, while the dome was built by the Allison Steel Company of Phoenix. Since astronomers would not be present in the dome while observing, special precautions were taken to protect the mountain personnel and equipment. An automatic weather station capable of closing the dome shutter and stowing the telescope in the event of high humidity, precipitation, or high winds was installed on the roof of the adjacent Administration

One of the necessary steps in the development of a Space Telescope was the successful operation of an earth telescope from a remote control point. This view shows the dome for the 50-inch Remote Control Telescope on Kitt Peak. The 50-inch telescope, once operated by an astronomer located nearly fifty miles away in Tucson, is still used nightly by observers operating from a control room within the dome. *(Photograph by the author)*

Building. An interlock circuit in the dome prevented movement of the telescope when the maintenance platform was not out of the way. An audio link to the system operator in Tucson monitored sounds in the dome, and a circuit was installed to lock out the Tucson remote control when individuals were working in the dome.[25]

The 50-inch aluminum mirror was completed in January of 1965 and the blank for the secondary mirror was received in the Optical Shop. When all was ready, the new Kanigen-coated aluminum mirror replaced the temporary 16-inch telescope. The lightweight mirror was designed to explore the center-supported thin shell (unribbed) concept advanced by Dr. Meinel. The Goddard Space Flight Center was considering such a mirror made of berylium for the second Orbiting Astronomical Observatory.[26]

Although the #1 36-inch telescope also had a metal mirror, its ribbed casting weighed only about one-third less than an equivalent glass blank and was mounted in a conventional cell. The new 50-inch was a true "lightweight" in the sense that it offered a considerable weight saving over glass or quartz. Designed by two Kitt Peak engineers, Frank Stuart and Mel Larson, there was no mirror cell in the ordinary sense. The mirror casting weighed 429 pounds and was quite thick at its center, while it tapered out to its periphery. The mirror was bolted at the center to a simple support bracket which in turn was bolted to the A-frames of the telescope tube. Together the mirror and support weighed only 738 pounds, compared to a glass blank which alone would have tipped the scales at over 1300 pounds.[27]

At the heart of the automatic telescope and data-acquisition system was a Packard Bell 250 computer, under the control of a paper-tape input. Astronomers using the system "observed" from Tucson, while the telescope, 55 miles distant, automatically acquired desired program objects and collected the desired data. A two-way communication system, designed by Astrodata Incorporated of Anaheim, California, connected the astronomer with his telescope while observing was in progress. The observer's office in Tucson, the 50-inch telescope, and the telecommunications system were popularly referred to by astronomers as the "Kitt Peak Range."

A new optical system was placed in the 50-inch on November 16, 1970. The lightweight aluminum mirror and support system were replaced with a Cer-Vit (ceramic-vitreous) primary. The primary mirror was lightweighted by the manufacturer, Owens-Illinois, to only 30% of the original, solid blank weight, making the total weight of the mirror and cell about 820 pounds. The new mirror was supported axially by an air-bag system and radially by a mercury tube in a lightweight cell.[28] The telescope, while no longer remotely operated from Tucson, is still heavily subscribed by astronomers.

The second major area of involvement for the Space Division was a true Kitt Peak Space Program. Utilizing the dependable Aerobee rocket, dozens of launches took place beginning with the launch of *Aerobee 1* on April 12, 1963 from White Sands Missile Range. Although this first launch went without a hitch, the nose cone failed to separate, thus preventing the collection of any data and also preventing the recovery package from operating.

Utilizing Aerobee rockets, the Kitt Peak Space Division performed dozens of launches from White Sands Missile Range in New Mexico. Various rocket developmental concepts were explored, including separating nose cones and different types of recovery packages. This photograph shows an Aerobee being readied for launch.

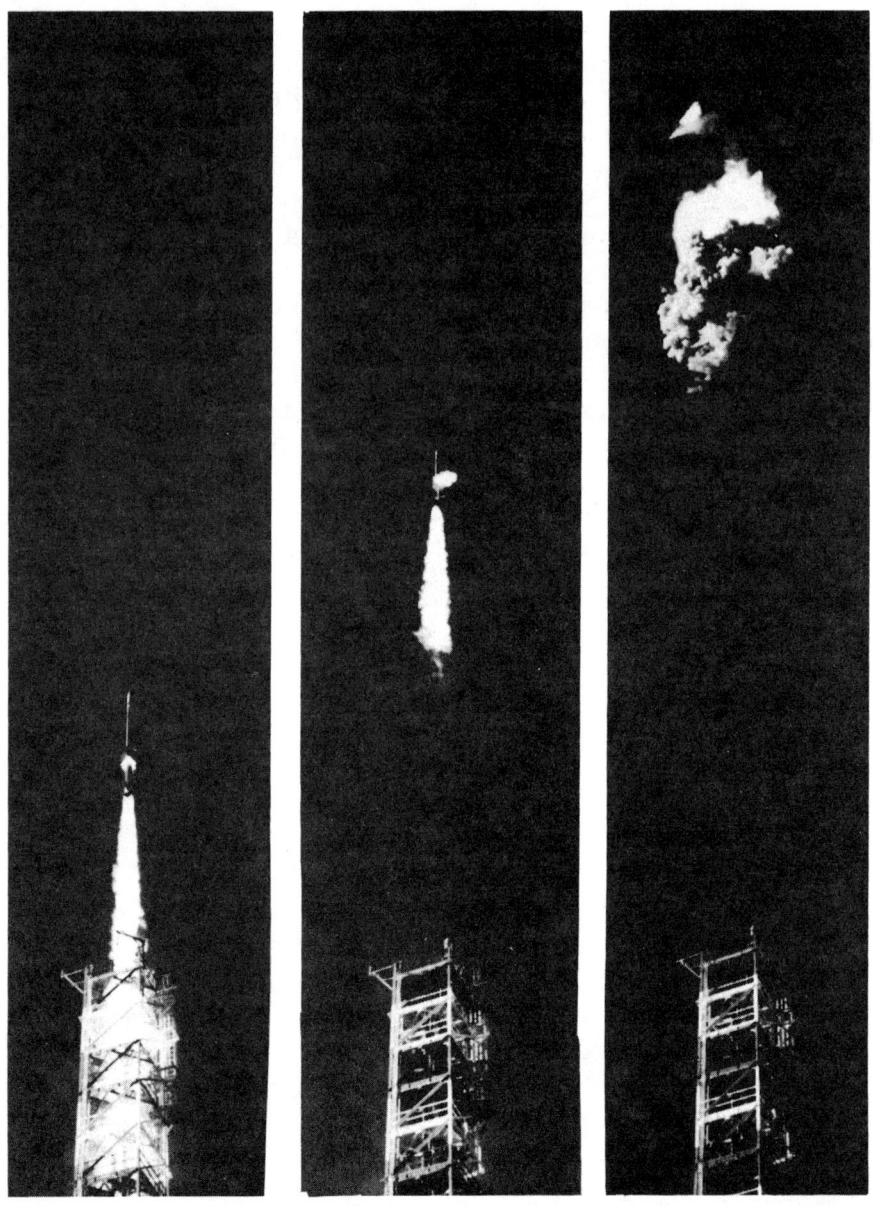

Moments after lift-off, the flight of an Aerobee rocket is cut short by a dramatic explosion. Despite several such failures, the Kitt Peak rocket program returned mountains of data on dayglow measurements and the zodiacal light.

The second Aerobee flight took place on June 25, 1963. The flight was termed successful although problems persisted with the nose cone separating properly. During the third Aerobee flight on November 4, 1963, the rocket performed ideally and the nose cone separated perfectly, but no data were collected due to instrument problems.

Various rocket developmental concepts were experimented with, from separating nose cones to various types of recovery packages. Of particular interest, the fourth Aerobee which was launched on April 7, 1964, did not even carry a parachute. The scientists hoped the payload would tumble slowly and gracefully to the ground, resulting in little or no damage to the instrument package. Unfortunately, unlike previous flights, the rocket did not break up on reentry into the earth's atmosphere; it came down like an arrow. The rocket with the payload still attached impacted the ground at an estimated 1000 to 5000 feet per second. The tail section barely protruded from the ground.

Experiments continued with instrument package designs, the Automatic Control System, and the Primacord (an explosive rope used to separate the payload from the rocket). Despite many heartbreaking failures (an Aerobee launched on January 18, 1966 was a complete failure; another rocket blew up on launch December 16, 1966; and as late as December 12, 1969, a malfunction destroyed a rocket and payload two seconds after launch), Kitt Peak scientists, technicians, and astronomers kept at their task. Many Aerobee flights were 100% successful, reaching altitudes of over 130 miles, and yielding astronomers a considerable amount of very useful scientific data. In addition to data on dayglow measurements and the zodiacal light, the technology gained from the rocket flights materially aided the United States space effort. Unfortunately, faced with ever--increasing operating costs and a diminishing budget, the observatory was forced to abandon its rocket program in July of 1973.

An offshoot of the Kitt Peak Space Division occurred during June of 1968 when astronomers successfully used the McMath Solar Telescope to direct a laser beam at the *Surveyor VII* landing site on the moon. Although six observatories made the attempt at bouncing laser beams off the lunar retro-reflectors, only Kitt Peak and one other observatory were successful.

Another offshoot of the Space Division was the "Airglow" Lab constructed on the mountain in 1963. The laboratory housed experiments concerned with the development of instrumentation for spectrophotometry at very low intensities. Studies were conducted of the twilight, dayglow, and night airglow emissions. The empty rectangular shell of the Airglow Lab, visible on the left, remains on the site, a vivid reminder of a very exciting past. On the right is the dome for the Burrell-Schmidt Telescope of Case Western Reserve University. *(Photograph by the author)*

The Nicholas U. Mayall 158-inch Telescope

We know that man's deepest wonder about himself and the world around him was first stirred as he turned his eyes toward the heavens and contemplated their vastness, their beauty, their orderliness, and the influence their changes seemed to have on his life....

...Let us hope that we never lose that endless need to know -- that spirit that keeps us forever reaching for the stars.[29]

Dr. H. Guyford Stever

In March of 1961, acting on a proposal by Dr. McMath, the AURA Board of Directors decided to begin design and location studies for a telescope having a mirror 158 inches in diameter. Because of its outstanding work on the McMath Solar Telescope, the Chicago firm of Skidmore, Owings, and Merrill was selected as architect for the initial studies of the dome and building. Four major engineering organizations were approached to prepare proposals for the preliminary design of the telescope. Westinghouse Electric Corporation of Sunnyvale, California, was finally selected to begin design studies for the telescope base, fork ring, and tube. Additionally, Westinghouse was to study potential mirror mounts, handling systems, bearings, and drive systems.

◄— The Nicholas U. Mayall 158-inch telescope. Weighing about 300 tons, the moving parts of the telescope are supported by a thin film of oil only 0.004 of an inch thick.

Was there a need for another large telescope? Dr. Ira S. Bowen, director of Mount Wilson and Palomar Observatories and a consultant to the AURA Board of Directors, at AURA's request, prepared the following statement:

On completion of the Palomar Observatory and initiation of joint operation with the Mount Wilson Observatory, a guest investigator program was initiated to permit astronomers from other institutions to use such telescope time as was not required by the Observatories' own staff. During the first decade of operation of the program, 80 astronomers from 41 institutions made 186 visits to the Observatories. Most of these visitors used our instruments to carry out the part of their programs requiring telescopes of large aperture.

In recent years the growth of our staff and the addition of Carnegie and N.S.F. fellows and graduate students from the rapidly expanding department of astronomy at the California Institute of Technology has made necessary the reservation of a larger and larger fraction of telescope time for the Observatories' own staff. It has become necessary to turn down an increasingly large fraction of the requests for telescope time by guest investigators, including some of the most distinguished astronomers of the country.

The above experience of these Observatories is but one indication of the rapidly growing interest in astronomy and of the fact that the number of astronomers wishing to carry out observational studies is rapidly outgrowing the telescope capacity of the country. Unless the future development of American astronomy is to be seriously hampered by the lack of instruments, it is very important that plans be started at the earliest possible moment to increase this capacity and especially that of instruments of large aperture.[30]

Yes, there was definitely a need for another large telescope, but how large should it be? The decision for a telescope incorporating a 158-inch diameter mirror was a compromise. The telescope had to be large enough to attack problems adequately at the frontier of astronomical research, and it also had to carry an efficient prime-focus observer's cage. However, to avoid a considerable amount of new experimental engineering, the telescope had to be either a copy of the 120-inch on Mount Hamilton or the 200-inch on Mount Palomar brought up to date, or an interpolation somewhere between these two designs. The telescope's cost had to be consistent with funds likely to be available from the National Science Foundation, and the telescope had to be ready within a reasonable length of time.

After elaborate site testing, the very summit of Kitt Peak was selected as the best possible location for the new 158-inch telescope. Site preparation was begun by AURA personnel in April of 1967 by blasting the southwest portion of the summit ridge level. Construction of the building and the massive telescope pier was begun in March of 1968 by the M. M. Sundt Construction Company of Tucson. Sundt was the prime contractor for the pier, building, and dome under a contract let on October 14, 1967.

The incredible task of pouring the concrete for the telescope pier was performed in July of 1968:

In continuous shifts, under blazing sun by day and under hanging construction lights by night, the construction workers poured the pier by the slip-form process, in which a mold is slowly raised as it is filled with concrete. By the time the mold leaves the bottom of the cylinder, the concrete there has already set. [31]

The monolithic pour required three days and three nights. Bill Daggett, Manager of Telescope Operations Support on the mountain, remarked that the construction lights used at night while the pour was taking place "made the construction site look like a very, very large birthday cake." The 37-foot diameter hollow cylinder rises 92 feet above the ground and, to prevent unwanted vibration from reaching the telescope, is completely isolated from the building. The 18-inch thick walls, extending down to bedrock, are reinforced with steel.

The unique appearance of the 158-inch telescope building is a result of using ten tubular frame modules called "hexahedrons." Each hexahedron was fabricated in Phoenix in one piece, trucked to the site, and erected at the rate of one per day. Each of these impressive modules stands nearly 96 feet tall, spans 33 feet, and tips the scales at a monstrous 35 tons. [32]

In a continuous operation lasting three days and three nights, the massive concrete pier for the 158-inch telescope was cast in July, 1968. This photograph taken on July 11 shows workers pouring the pier using the slip-form process in which a mold is slowly raised as it is filled with concrete. Bill Daggett, Manager of Telescope Operations Support on the mountain, remarked that the construction lights used during the three nights the pier was being cast "made the construction site look like a very, very large birthday cake." Astronomers observing with other telescopes on the mountain were less than thrilled about the interference caused by the lights.

When this photograph was taken on September 5, 1968, five of the ten hexahedrons which comprise the structural framework for the 158-inch building had been installed. Each of these tubular frame supports stood nearly 96 feet high, spanned 33 feet, and weighed 35 tons. The hexahedrons were erected at the rate of one per day.

Appearing much like a monstrous spider web, the steel ribs for the 158-inch dome were placed in April of 1969. Prior to erection of the dome, the lower section of the telescope base frame was installed under the main floor of the observing level. This photograph taken on April 18 shows construction workers more than 100 feet above the ground precariously fastening the ribs in place.

Once installation of the hexahedrons was complete the 14-foot high drum girder was erected atop the 10-sided polygonal structure. One of the sections of the drum girder weighed 45 tons, the heaviest single piece of material to be handled during the construction job.[33] By the end of February 1969, the dome trucks and ring girder of the dome were being installed. Building progress had advanced sufficiently so that the most popular conditioning exercise for visiting astronomers and the resident mountain staff was climbing the many flights of stairs to the drum girder level.

Prior to erection of the steel dome, the lower part of the telescope base frame was delivered by the Everett, Washington, Heavy Machinery Division of the Western Gear Corporation and installed under the main floor of the observing level, about 100 feet above the ground. Weighing nearly 500 tons, the dome rotates on its 32 trucks in 5 1/2 minutes. The shutter — more than 27 feet wide — weighs about 30,000 pounds. Being an "up and over" type, the shutter will fully open or close in 2 1/2 minutes.

The dome, built to withstand hurricane force winds of 120 miles per hour, incorporates a double shell structure, having an outer diameter of 108 feet, and an inner diameter of 102 feet. The exterior of the dome consists of 1/4-inch steel cover plates welded to the dome superstructure. The interior of the dome is covered with 26,600 square feet of embossed aluminum insulating panels.

The 158-inch building was completed by the contractor in late 1970. Standing 187 feet high, the building contains approximately 30,000 square feet of floor space. The interior is divided into offices, workshops, darkrooms, sleeping quarters, and storage areas.

The concept for the 158-inch telescope mounting was developed by W. W. Baustian, chief engineer for Kitt Peak, and is a modification of types originally investigated for the 200-inch, Mount Palomar telescope. In many ways the mounting is similar to the 200-inch, but the "horseshoe" is located at the declination axis. This allows the telescope tube to swing freely between the tines of the yoke, rather than between the struts running from the yoke to the south bearing. The moving parts of the mounting weigh about 300 tons, and turn on a thin film of oil only 0.004 of an inch thick on eight hydrostatic bearings. The telescope is so precisely balanced it can be moved by a 1/2 horsepower motor.

The telescope moves on two sets of main support bearings; the "horseshoe" bearing for right ascension (corresponding to longitude in the sky) and a perpendicular set of bearings for declination (corresponding to latitude in the sky). The right ascension bearings are mounted parallel to the earth's axis of rotation and the telescope can track an object westward across the sky by moving on these bearings at the rate necessary to

When installation of the steel ribs for the 158-inch dome had been completed, the tedious task of enclosing the dome began. The exterior of the dome was covered with ¼-inch thick steel plates welded to the dome superstructure. The interior of the dome was covered with embossed aluminum insulating panels two inches thick. The dome was built to withstand hurricane force winds of 120 miles per hour.

When this photograph was taken on August 20, 1969, construction of the 158-inch dome was nearly complete. The dome, weighing nearly a million pounds, rotates on 32 sets of wheels which follow a circular rail track. The shutter, not visible, is more than 27 feet wide and weighs 30,000 pounds.

Standing 187 feet high, the 158-inch building dominates the summit of Kitt Peak. Containing 30,000 square feet of floor space, the interior of the building is divided into offices, workshops, darkrooms, sleeping quarters, and storage areas. *(Photograph by the author)*

On October 14, 1971, when this photograph was taken, work was progressing well on the 158-inch telescope. The mounting, designed by W. W. Baustian, is a modification of types investigated for the Mount Palomar 200-inch telescope, and in many ways resembles that instrument.

compensate for the earth's rotation. As the telescope moves, the dome automatically rotates to allow the telescope a clear view of the sky through the slit. All motions of the telescope and dome are controlled by a computer.

The yoke is approximately 41 feet in diameter, large enough to contain the tube and still have sufficient structural strength to hold deflections within tolerable limits. The yoke acts as the north journal of the polar axis, as in the 200-inch Mount Palomar telescope. A pair of radial oil pads supports this north journal, and pairs of radial and axial thrust oil pads are at the south end of the polar axis. A large triangular base frame supports the entire instrument. To facilitate critical alignment of the polar axis, the frame is mounted on a south pivot and two adjustable north supports. The total weight of the mounting is a very massive 706,600 pounds.

95

Drawing of the 158-inch building. *(Courtesy of Kitt Peak National Observatory)*

Workers assemble the 33 pneumatic pads which support the 15-ton mirror of the 158-inch telescope. The support system prevents the figure of the mirror from changing as the telescope is pointed to different locations in the sky.

The primary mirror rests in a rigid cell with its back supported by 33 analog computer controlled pneumatic pads. The edge of the 158-inch glass disk is supported by 24 counterbalanced push-pull mechanical lever supports. The support system maintains the mirror in a supported condition such that the figure of the mirror is not disturbed when the telescope is pointed to different locations in the sky (different orientations relative to gravity).*

At the April 15, 1963 annual meeting, the AURA Board of Directors approved fused quartz as the material for the new 158-inch blank. Fused quartz was selected instead of pyrex (which had been used for the 84-inch mirror) because of its lower sensitivity to temperature changes. Also, a disk of fused quartz would be easier to grind, polish, and figure in the Kitt Peak Op-

*See Reference Note 32.

97

tical Shop. New processes developed by several manufacturers permitted quartz to be fused in larger sizes not attainable in the past.* The 158-inch blank would be the largest blank yet made of this material.

On December 31, 1964, the General Electric Company was awarded a contract in the amount of $1,150,000 for the 158-inch blank. General Electric had pioneered the development of fused quartz blanks at their Lynn plant in 1931 in an attempt to build up to a 200-inch blank for the Palomar telescope. The technique employed at that time involved a spray process which laid down and melted highly pure silica particles on a base. In this manner a monolithic piece was built up, layer by layer. While the 200-inch blank would have been too expensive, and Corning Glass Works was given an order for a disk of pyrex, General Electric did achieve a number of small and intermediate size quartz disks. Most notable were the two 63-inch diameter mirrors which temporarily served in the McMath Solar Telescope.

In the case of the new 158-inch blank, new technology attained at General Electric permitted fusing a number of quartz ingots together, thus forming a disk of the required dimensions. To form the individual ingots, high-purity silica sand was placed in crucibles and fused in a vacuum furnace at temperatures approaching 3300°F. After fusion, the ingots were ground and fitted together to form the 158-inch blank. The ingots were then fused together in a larger furnace (not in a vacuum) to form the finished blank.

The 158-inch disk consisted of three layers of hexagonally shaped quartz ingots. The substrate was composed of two layers of ingots 12 inches high and 6 inches across. Capping this substrate was a layer of ingots 6 inches high and 20 inches across. Four fusions were required in melting together the individual ingots. After the first fusion of the substrate there were multiple cracks throughout the blank. Changes were made in the refractory materials, and again the great disk was super-heated. After the second fusion one large crack persisted in the blank, successfully remedied in a third fusion. The fourth and final fusion added the 6-inch cap to the now monolithic base and was successful in producing a blank of the required thickness.

---

*See Reference Note 32.

The AURA inspection group met at the Lamp Glass Division of General Electric in Cleveland, Ohio, on September 28, 1967, for acceptance tests of the blank. Test results for strain and bubble content were better than the specifications required in all parameters, and the blank was officially accepted. Late in October of 1967 the million-dollar disk began its 1800-mile rail journey to Tucson. The blank stood upright in a cushioned cradle on an under-frame flatcar, the same car carried a large "spreader" nearly 19 feet long to facilitate unloading the blank in Tucson. On October 30 the long ride from Ohio to Arizona was completed.

Once the blank had arrived at the Kitt Peak Optical Shop, opticians began the sensitive task of grinding, polishing, and figuring the 15-ton disk in a painstaking process requiring three years. The largest blank previously worked in the Optical Shop was the 84-inch mirror. Because the grinding machine would not

Optician Harold Wirth lowering the grinding tool onto the 158-inch glass blank in the Kitt Peak Optical Shop. *(Photograph, March 13, 1969)*

99

accommodate the larger 158-inch disk, an AURA-designed, special large grinding machine was installed in August of 1967.

After fine grinding was completed, a general clean-up of the grinding room and equipment was performed. Painters were called in to repaint the entire room, including the overhead beams, walls, and the grinding machine. All the heavy grit was therefore either washed away or painted down, and only the finest abrasives and polishing compounds were permitted in the area. Opticians and visitors entering the polishing room were required to wear clean plastic boots or other clean footwear.*
Only after the disk had been polished within an accuracy of one-millionth of an inch was a reflective coating of aluminum, 1/1000 the thickness of a human hair, deposited on the front surface.

Over a period of three years the Kitt Park opticians carefully ground, polished, and figured the 15-ton 158-inch disk.

With the letting in of "first light" achieved on February 27, 1973, the 158-inch telescope became the world's second largest operating optical telescope. When the ten-million-dollar project was formally dedicated on June 20, 1973, the telescope was officially named after Dr. Nicholas U. Mayall, who served as Kitt Peak's director from 1960 until his retirement in 1970. The keynote speaker at the dedication ceremony, Dr. H. Guyford Stever, director of the National Science Foundation, said in his dedication address:

We know that man's deepest wonder about himself and the world around him was first stirred as he turned his eyes toward the heavens and contemplated their vastness, their beauty, their orderliness, and the influence their changes seemed to have on his life. From this initial curiosity, and the investigation and knowledge to follow, was to emerge so much that would bring order and meaning and new growth to our lives. What we have dedicated here today is not only a new telescope to carry out the goals of a great observatory. It is a symbol of man's continued determination to strive for more fundamental knowledge — to search out the limits of the universe, to probe for an understanding of the very essence of matter and life, to attempt always to know the unknown. That is the meaning of basic research — but more than that, it is one of the bases of our humanity. Let us hope that we never lose that endless need to know — that spirit that keeps us forever reaching for the stars.*

---

*See Reference Note 29.

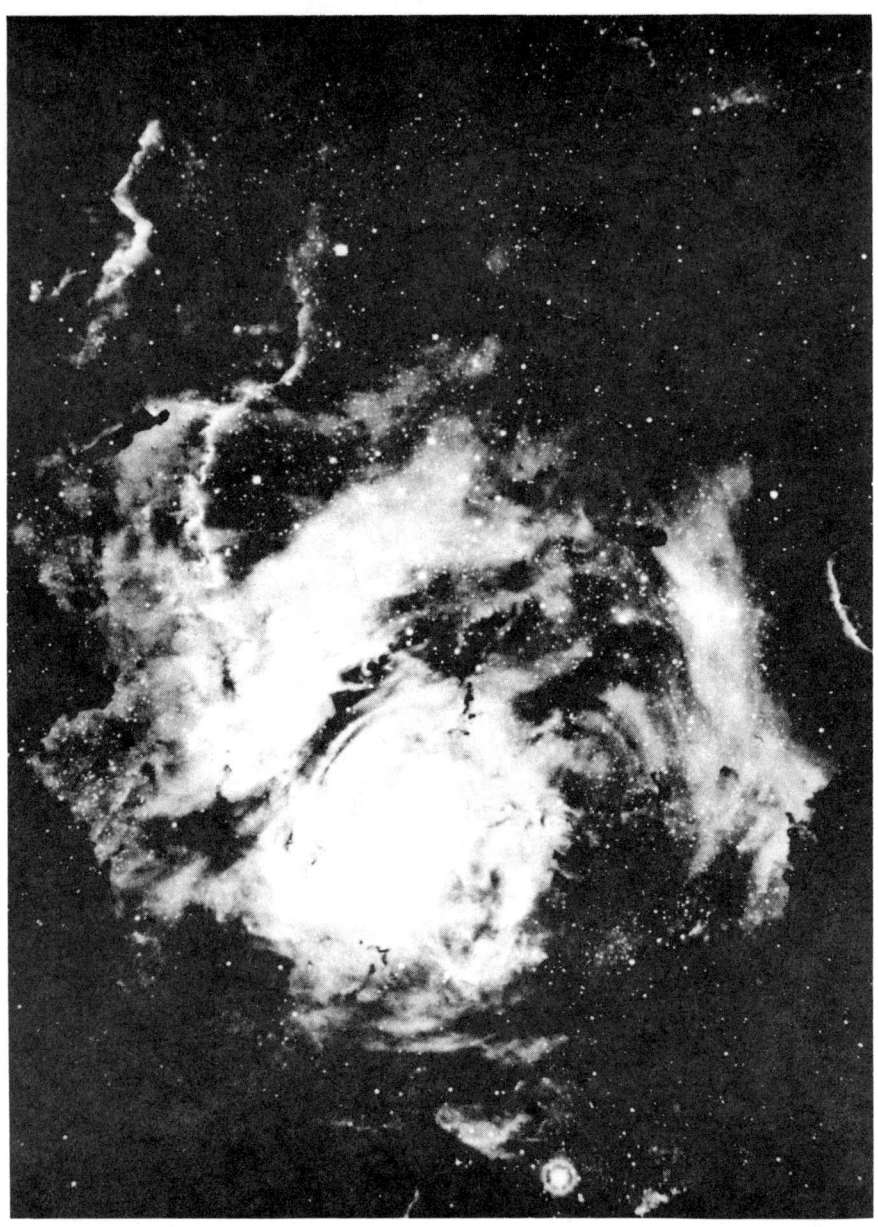

The majestic Lagoon Nebula, NGC 6523 in the constellation Sagittarius, is one of the finest diffuse nebulae visible in our sky. The intricate detail between the mixed bright and dark nebulosity may be seen even in small telescopes. The light which formed this photograph had been traveling through space for over 5000 years. This photograph was taken with the 158-inch telescope.

The spiral structure of the beautiful Whirlpool Galaxy, NGC 5194 in the constellation of Canes Venatici, was first detected by Lord Rosse at Parsonstown, Ireland, with his 6-foot reflector in 1845. Such "spiral nebulae" were at first believed to be new solar systems within our own galaxy in the process of formation. In 1923 Astronomer Edwin Hubble proved that these objects were actually galaxies composed of billions of stars. The Whirlpool Galaxy is an incredible 35 million light years from Earth, yet is one of the nearest galaxies. This photograph was taken with the 158-inch telescope.

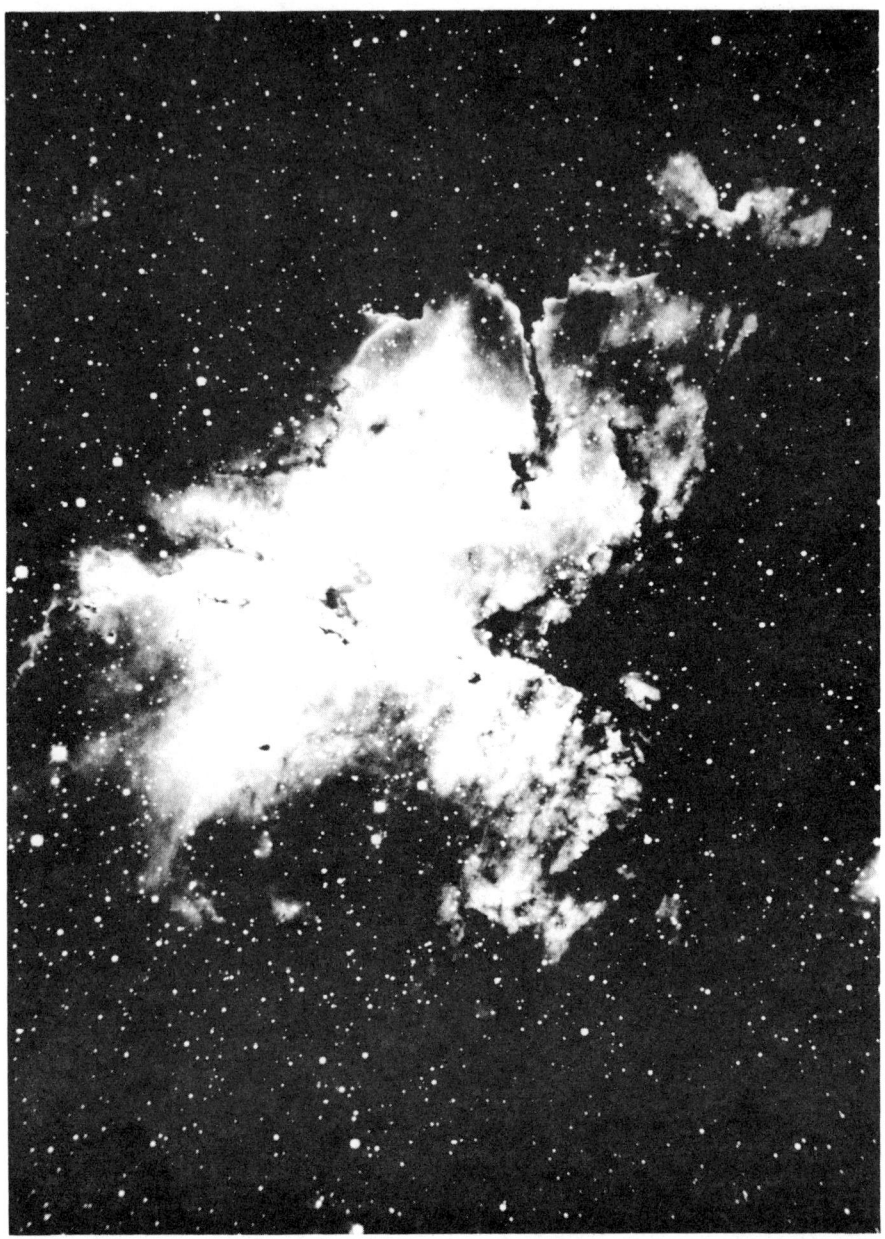

The stately Eagle Nebula, NGC 6611 in the constellation Serpens, is a vast diffuse nebula which contains a large scattered star cluster. This stellar nursery lies about 8000 light years distant in the Sagittarius spiral arm of our galaxy. This photograph was taken with the 158-inch telescope.

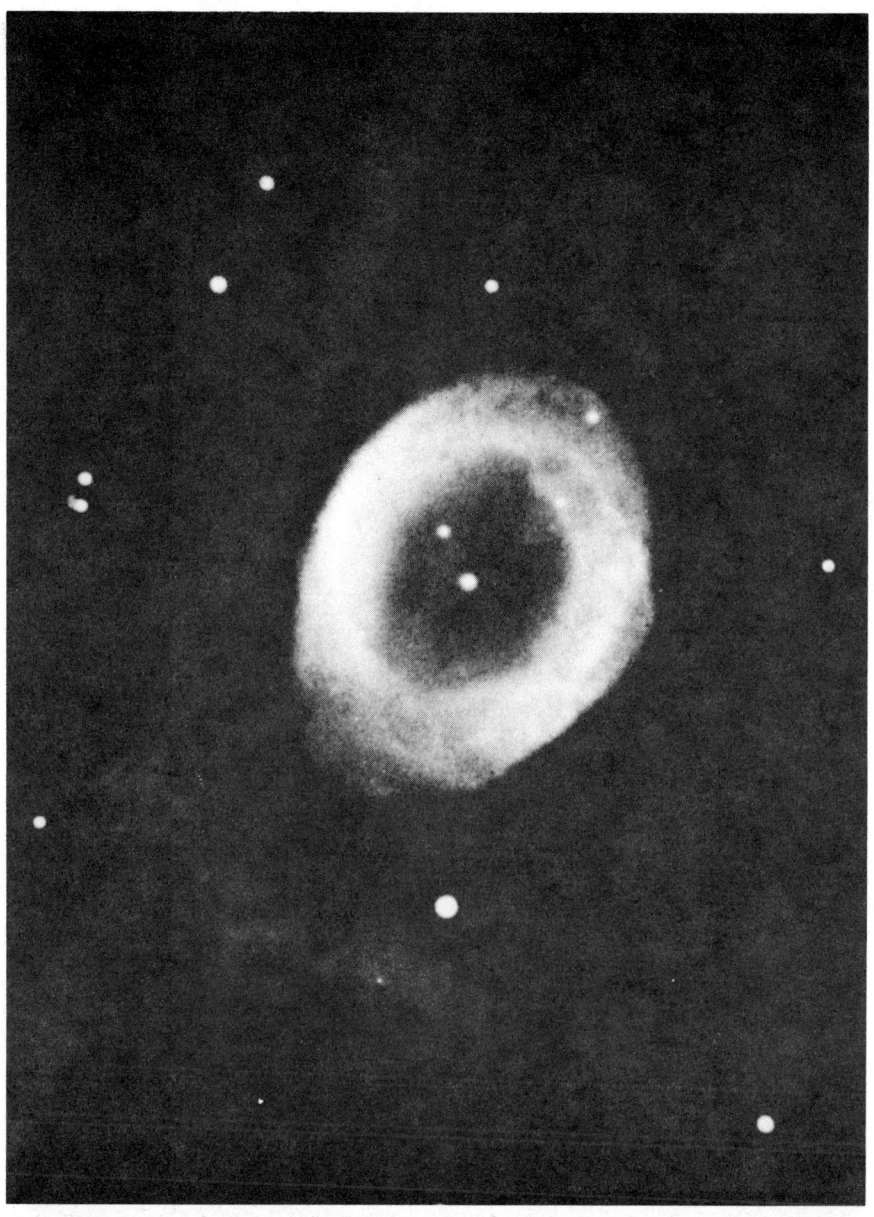

The Ring nebula, NGC 6720 in the constellation Lyra, was the first planetary nebula to be discovered. Planetary nebulae form when an aging star (for reasons not thoroughly understood) ejects its outer layer of gases. The resulting shell of gas slowly expands into space; in the case of NGC 6720 the shell is expanding at 12 miles per second. This photograph was taken with the 158-inch telescope.

105

The 158-inch telescope was used to record this photograph of the globular cluster NGC 6341 in the constellation Hercules. Globular clusters appear as tightly packed balls of one hundred thousand stars or more. NGC 6341 lies within the confines of our own Milky Way Galaxy, being about 30,000 light years distant.

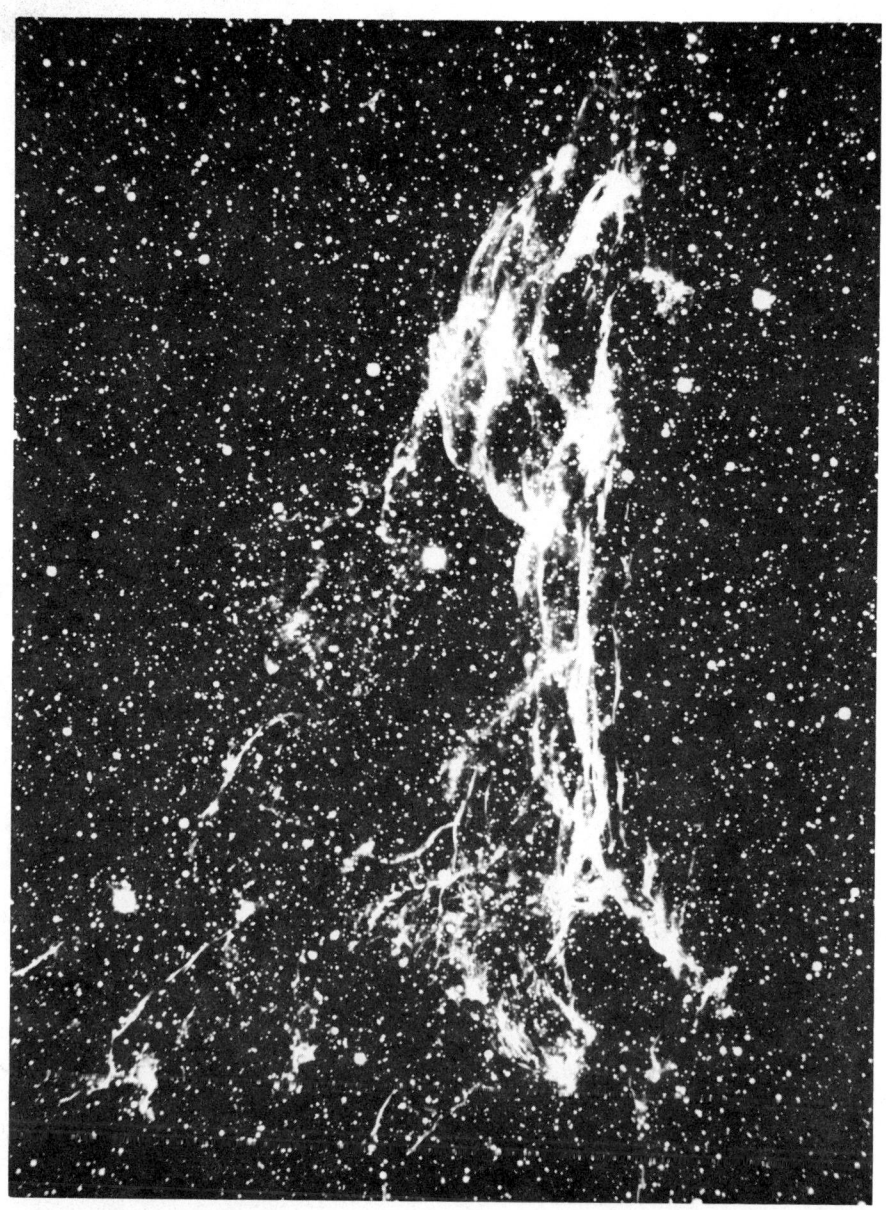

The Veil Nebula, NGC 6992 in the constellation Cygnus, was discovered by Sir William Herschel with his 18-inch telescope in 1784. Some astronomers believe these delicate filaments of gas are the remnants of a great stellar explosion which probably occurred more than thirty thousand years ago. This photograph of the central portion of the Veil Nebula was taken with the 158-inch telescope.

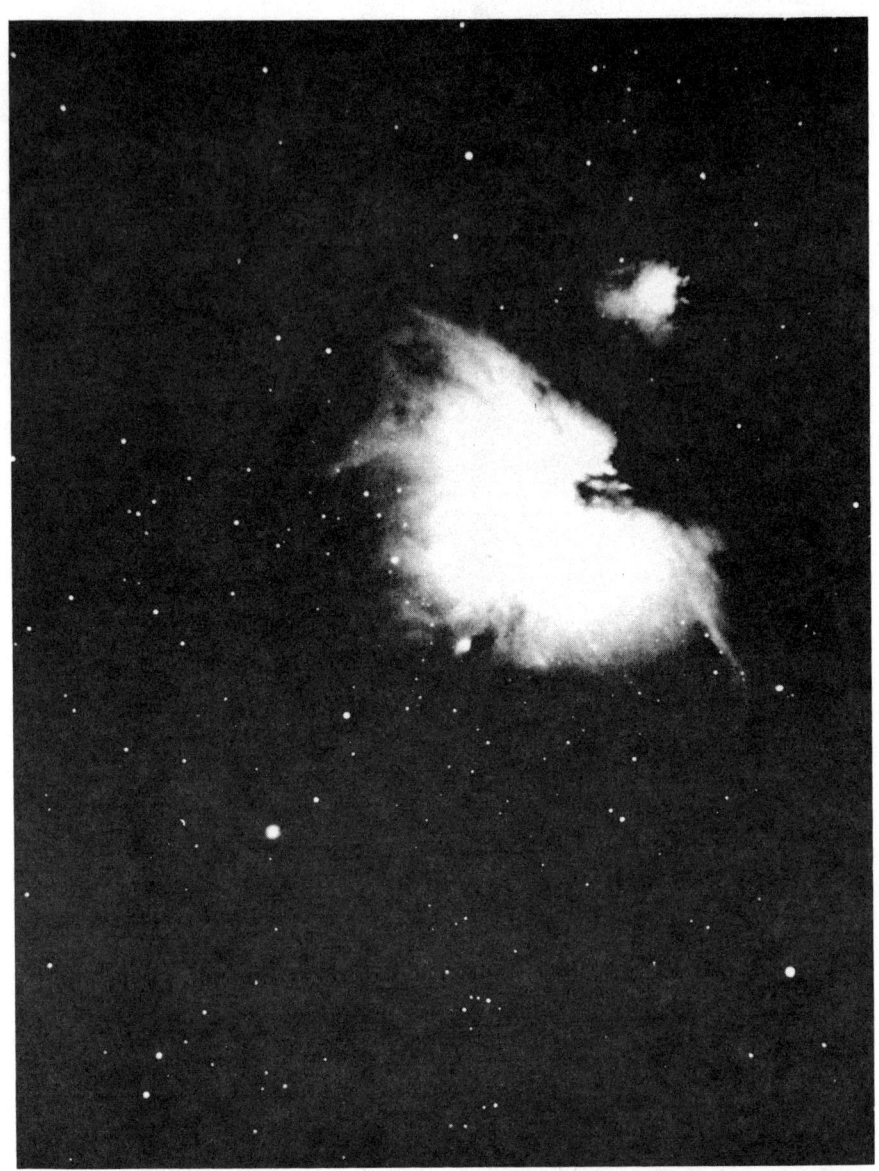

The Great Nebula in Orion, NGC 1976, is one of the finest examples of a diffuse nebula visible in our sky. This vast cloud of gas and dust lies nearly 1700 light years from the earth and is roughly 30 light years across — more than 20,000 times the diameter of our solar system. Within the turbulent confines of this nebula astronomers have found a stellar birthplace as new stars are being formed from the nebula's material. This photograph was taken with the 158-inch telescope.

The 158-inch telescope dome dominates this view. Other domes are (left to right): University of Arizona 36-inch; #2 16-inch; Belton-Schmidt; University of Arizona 90-inch; and the 50-inch. *(Photograph by the author, 1977)*

Other Astronomical Facilities

The observatory grew larger and larger in both facilities, equipment and personel. Many institutions participated in the growth and benefits of this National Astronomical Observatory.

Many other astronomical facilities were added over a period of time to the main observatories. Among them are: the Burrell-Schmidt Telescope, moved from Case Western Reserve University to the mountain in 1979; the McGraw-Hill Observatory for which construction began in 1974; the University of Arizona 36-inch and 90-inch telescopes; and the National Radio Astronomy Observatory which became operational in 1966.

Late in 1974 construction began on the southwest ridge of Kitt Peak for the new McGraw-Hill Observatory. The observatory was to house the University of Michigan 52-inch reflector, originally located in Stinchfield Woods about 20 miles outside Ann Arbor, Michigan. Because of the poor weather and observing conditions at the Michigan site, it was decided to relocate the telescope to the Southwest. A consortium of three institutions, the University of Michigan, Dartmouth College, and Massachusetts Institute of Technology, was formed to pay the operating costs of the observatory, and support came from the McGraw-Hill Company and the Alfred P. Sloan Foundation for the new buildings.

◄— The Burrell-Schmidt Telescope of Case Western Reserve University was moved to Kitt Peak in 1979. The telescope can be used either for conventional Schmidt photographs or objective-prism photographs.

111

The 84-inch dome is in the center. The small dome to the immediate left of the 84-inch dome houses the #3 16-inch telescope. The 50-inch dome lies to the left adjacent to the administration building. The large dome to the lower right houses the Burrell-Schmidt Telescope, which was moved to the mountain in 1979. The small dome at the lower right was the first dome erected on the mountain and is no longer used. To its left is a dome which was used to house a small Schmidt telescope. *(Photograph by the author)*

McGraw Hill Observatory.*(Photograph by the author)*

112

The observatory consists of two buildings. One houses the telescope and is topped with a 30-foot diameter gear-driven dome. The second building houses living quarters, shops, and the computer. The telescope, which was designed by Tinsley Laboratories of Berkeley, California, was installed and operational only six months after the ground-breaking ceremony. Formal dedication of the observatory took place on November 15, 1975.

The 36-inch telescope of the University of Arizona, originally erected on the Tucson campus in 1922, is now housed within a dome on Kitt Peak. This instrument was used to show representatives of the Papago Tribal Council the moon and planets in an effort to obtain permission to construct the National Astronomical Observatory atop Kitt Peak. City lights and smoke

Dome housing the University of Arizona 36-inch telescope. *(Photograph by the author)*

hampered the telescope's usefulness in Tucson, so it was decided to sublease land on Kitt Peak from the National Science Foundation and relocate the telescope. The building was finished and the telescope was moved up the mountain in late 1962.

In the summer of 1965 the National Science Foundation granted $1,300,000 for construction of the University of Arizona's 90-inch telescope and its optics. The State of Arizona gave an additional $700,000 for the building and dome. Construction of the 70-ton telescope began in March of 1966 at the Boller

Building and dome for the University of Arizona 90-inch telescope. *(Photograph by the author)*

& Chivens plant while the Lamp Glass Division of General Electric prepared to cast the 7000-pound primary mirror blank. Over 250 hexagonal ingots of fused quartz were fused together at 2000°C. The finished blank was figured at the University of Arizona Optical Sciences Center and aluminized at Lick Observatory.

In contrast to the conventional hemispheric dome, the 90-inch has a quarter cylinder set horizontally at the top of the observatory tower. Such a design permits use of all-straight structural-steel members, instead of curved ones, in the dome construction. When opened for observing, the biparting shutters move horizontally onto the flanks of the quarter cylinder, where they are better protected from the wind than if on a hemispherical dome. The 37-foot-high pier is also unconventional, having an inner and outer shell of concrete blocks 8 inches thick, with concrete poured in between. This design, with a resulting wall thickness of nearly 2 feet, is less costly than using slip forms, particularly on a mountain top. The reflector was formally dedicated on June 23, 1969, only four years after the project was begun.[34]

National Radio Astronomy Observatory. *(Photograph by the author)*

Perched atop Horseshoe Ridge, the National Radio Astronomy Observatory began studying the heavens in 1966. More than thirty different types of molecules have been detected in the space between stars; most of these molecules were discovered with this 36-foot radio telescope. The staff informally named their site "Caboodle Knob" so the whole "Kitt and Caboodle" would be in the Quinlan Mountains.

The millimeter-wave antenna of the National Radio Astronomy Observatory pointing to the zenith within its canvas-covered dome. After arriving at the base of the mountain on February 17, 1966, the antenna was kept under guard until conditions were ready at the site. On May 26 the 36-foot-diameter antenna was moved up the mountain to its site located on Horseshoe Ridge about one mile southwest of the summit.

116

Maintenance

The lofty mountain top hums with activity as routine maintenance is performed on the telescopes, computers, and equipment.

The stars have barely faded from view in the dawn of a new day as a technician threads his way to the many telescopes located on Kitt Peak. He pauses just long enough in each dome to assure himself that the dome slit is securely closed and the telescope with its sophisticated instrumentation is properly powered down and stowed. The telescope maintenance log — on which the night's observer notes any problems or peculiarities with the equipment — is picked up from each dome.

After a brief stop to see that the solar telescopes and equipment are all functioning smoothly, the technician heads to his office where the telescope maintenance logs are carefully reviewed. Appropriate observatory personnel are notified of any problems, needed repairs, or special requests noted by the astronomers. Next, the telescope observing schedule is examined to determine which instruments must be changed and prepared for the next observer.

The day progresses at a feverish pace as some instruments are removed from telescopes and others installed. The telescopes are then carefully balanced to ensure smooth trouble-free operation. Each instrument is meticulously checked and rechecked to ensure proper operation during the forthcoming night. Problems with instruments and equipment must be analyzed and solved during the day, rather than wasting precious observing time at night.

The lofty mountain top hums with activity as routine maintenance is performed on the telescopes, computers, and equipment. There are lenses and mirrors to be cleaned, gears to be greased, walls to be painted, and floors to be scrubbed. There are power supplies to check, spare parts to be repaired, and supply inventories to be maintained. Rarely is there a dull moment on the mountain.

Additionally, there are frequent training sessions to be attended by the entire mountain staff. When you are situated on a remote mountain top with the nearest hospitals and fire departments located 55 long miles away, emergencies can not wait for outside assistance. Therefore, the mountain-based staff is trained in such emergency procedures as handling the observatory's fire truck and fire-fighting equipment, and driving the observatory's ambulance. First-aid procedures such as treating rattlesnake bites, scorpion stings, and sprained or broken limbs, are also covered. Thus, Kitt Peak personnel are trained and equipped to handle most emergencies which might arise.

Astronomer Don Hayes at a computer terminal preparing for a night's observing run on the #1 36-inch telescope. Mounted on the telescope is an automatic guider, spectrograph, and reticon scanner. To the left of Don appears the telescope control console and power supplies. (A recent photograph by the author)

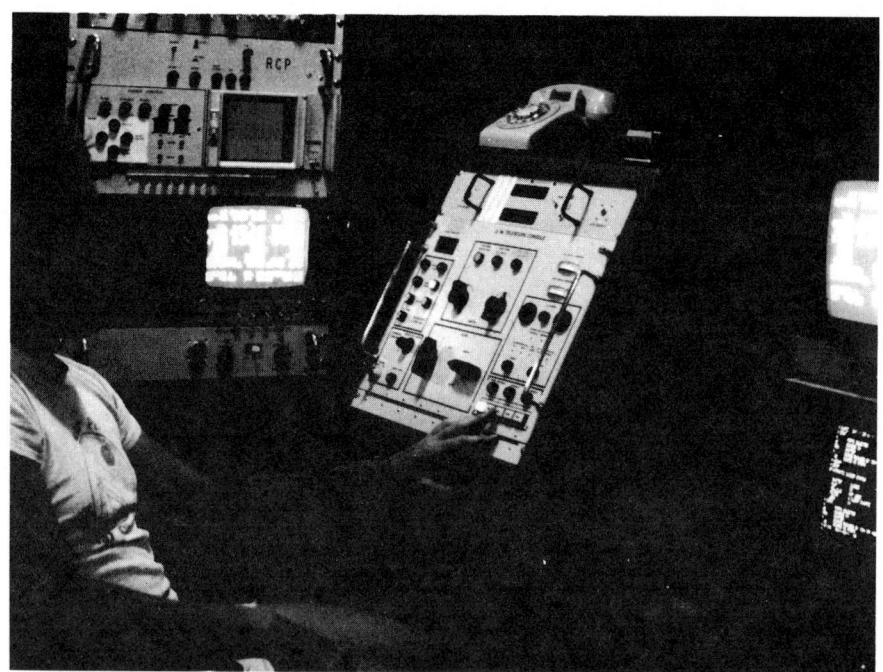

Technician Dean Ketelsen at the control panel for the 84-inch telescope. The complicated task of positioning the telescope is the responsibility of highly trained technicians and telescope operators. *(A recent photograph by the author)*

Use of the Observatory

It is the time of the "long eyes", the astronomers of Kitt Peak National Observatory.... For them the coming of darkness ushers in still more wonders....[35]

John M. Scott, Arizona Highways

Sometime during the afternoon the astronomers arise from their well-earned slumber. From points scattered across the face of the earth these astronomers have come to Kitt Peak to use the telescopes to collect their data. Preparations for their observing runs had begun months in advance. First, requests for observing time on the telescopes had to be submitted. All potential observers had to outline their proposed observing projects, as they vied for time on the great telescopes at Kitt Peak National Observatory. A panel of astronomers judges the requests, and only the most meritorious observing projects are approved. There simply is not enough available time on the Kitt Peak telescopes to satisfy the overwhelming demand. For example, three of every four astronomers requesting time on the 158-inch telescope must be refused.

Those fortunate enough to be granted time on the telescopes must immediately begin to prepare for their observing run, still months away. Each visiting astronomer is assigned a Kitt Peak staff astronomer who is intimately acquainted with the particular instrument to be used for the observing run, and correspondence is begun.

Every research project is different, placing unique demands upon the instrument and facilities. Although the same instrument may be used for several consecutive astronomer's runs,

The sprawling mountain facilities can be seen in this aerial photograph. The 158-inch dome appears in the upper right corner, the 84-inch appears to the left, and the McMath Solar Telescope is in the lower left corner.

adjustments to the instrument will no doubt be required. Different gratings may be required to study different regions of a star's spectrum, different photographic emulsions may be needed, and different computer programs may be required for each astronomer. The Kitt Peak staff astronomer works with the visiting astronomer to ensure that proper instrument and support equipment requirements are met for the observing run.

After a hasty dinner in the mountain cafeteria, the astronomers head for the telescopes to begin the night's labors. As the sun slowly sets in the west, steaming hot coffee is poured, dome slits are silently opened, telescopes are pointed, and observing begins in earnest.

If John Quincy Adams, who pleaded before Congress in 1825 for "lighthouses of the sky," could see the sprawling facilities of the Kitt Peak National Observatory, surely he would stare in

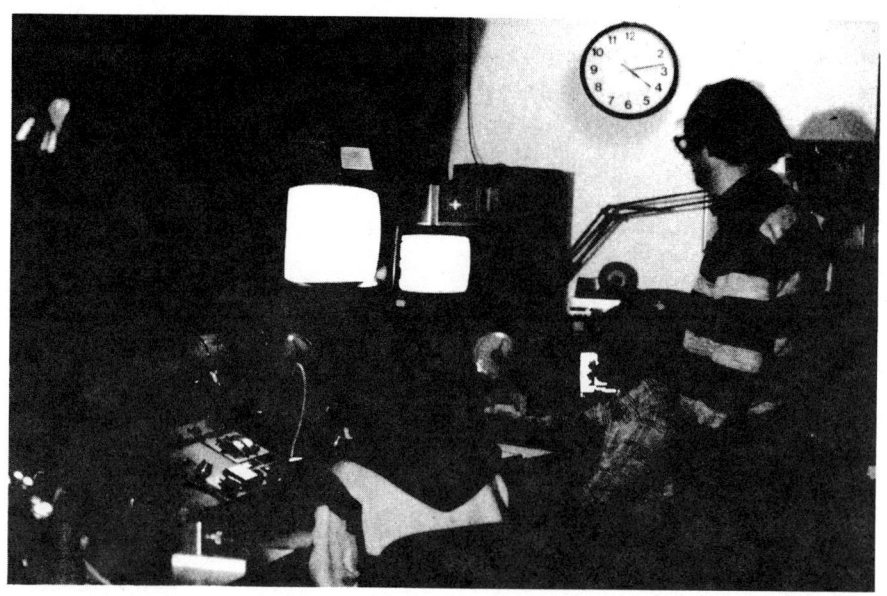

Astronomers Jean Goad (left) and Jay Gallagher (standing) while observing on the 158-inch telescope. Daryl Willmarth is seated at the myriad controls for the sophisticated telescope. *(A recent photograph by the author)*

Telescopes and astronomers at work. *(Photograph by the author)*

123

awe. Since 1958 astronomers have been traveling from across the globe to this superb observatory in an effort to unravel the mysteries of the universe. The astronomers are still coming, and the mysteries are still being unraveled.

Reference Notes

1. Irwin, John B., editor, *Proceedings of the National Science Foundation Astronomical Photoelectric Conference* (held at Lowell Observatory, Flagstaff, Arizona, August 31 to September 1, 1953), published by Indiana University, pp. 107-108.

2. Irwin, John B., editor, *Proceedings of the National Science Foundation Astronomical Photoelectric Conference* (held at Lowell Observatory, Flagstaff, Arizona, August 31 to September 1, 1953), published by Indiana University, p. 119.

3. Whitford, A.E. "The Plan for a New American Observatory," *Astronomical Society of the Pacific*, 68, pp. 115-117.

4. Bell, Trudy "Kitt Peak National Observatory: 'Telescope National Forest'," *Star & Sky*, 1, 4 (April 1979), p. 16.

5. "A New American Observatory," *Sky & Telescope*, 15, 3 (January 1956), pp. 107, 111.

6. Meinel, A.B. and Abt, H.A. "Preliminary Site Reconnaisance Report — National Astronomical Observatory Technical Report #2" (June 1955), p. 1.

7. Meinel, A.B. "The National Observatory at Kitt Peak," *Sky & Telescope*, 17, 10 (August 1958), pp. 493-498.

8. *Kitt Peak National Observatory* (a publicity booklet issued by the observatory in March 1960).

9. Carroll, John Alexander, editor, *Pioneering in Arizona — The Reminiscences of Emerson Oliver Stratton and Edith Stratton Kitt*, Tucson, Arizona: Arizona Pioneers' Historical Society, 1964, p. 142.

10. Clark, Geoffrey Anderson "Notes on the Kitt Peak Cave Cache, South-Central Arizona" (a paper written for the Department of Anthropology, University of Arizona), 1966.

11. Livingston, W.C. *Some Local Walks around Tucson and at Kitt Peak*, (December 1973), unpublished Manuscript.

12. Meinel, A.B. "Development Report for Kitt Peak — National Astronomical Observatory Technical Report #11" (April 1957).

13. "Kitt Peak National Observatory Monthly Report" (November 1959), p. 3.

14. "Kitt Peak National Observatory Monthly Report" (December 1959), p. 2.

15. "Kitt Peak National Observatory Monthly Report" (March 1964), pp. 3-4.

16. "Kitt Peak National Observatory Monthly Report" (June 1961), pp. 4-5.

17. "Kitt Peak's 80-inch Stellar Telescope," *Sky & Telescope, 23*, 1 (January 1962), pp. 4-9.

18. McMath, Robert R. "The Large Solar Telescope at Kitt Peak — I," *Sky & Telescope, 20*, 2 (August 1960), pp. 64-67.

19. McMath, Robert R. "The Large Solar Telescope at Kitt Peak — II," *Sky & Telescope, 20*, 3 (September 1960), pp. 132-135.

20. Wright, Helen *Palomar — The World's Largest Telescope*, Mac-Millan Company, New York, 1953, pp. 100-101.

21. Woodbury, David O. *The Glass Giant of Palomar*, Dodd, Mead, and Company, New York, 1966, p. 139.

22. Woodbury, David O. *The Glass Giant of Palomar*, Dodd, Mead, and Company, New York, 1966, p. 140.

23. Petrie, R.M. "Builder of Solar Observatories," *Sky & Telescope, 23*, 4 (April 1962), pp. 187-190.

24. "Kitt Peak National Observatory Monthly Report" (January 1963), pp. 16-18.

25. "Kitt Peak National Observatory Monthly Report" (December 1965), pp. 4-7.

26. "Kitt Peak National Observatory Monthly Report" (February 1962), p. 8.

27. "American Astronomers Report," *Sky & Telescope*, *34*, 4 (October 1967), p. 213.

28. "Kitt Peak National Observatory Monthly Report" (October-December 1970), p. 3.

29. Aebischer, Edward D. "Kitt Peak National Observatory," *Arizona Highways* (November 1973), p. 34.

30. "Kitt Peak National Observatory Monthly Report" (March 1962), p. 3.

31. Bell, Trudy "Kitt Peak National Observatory: 'Telescope National Forest'," *Star & Sky*, *1*, 4 (April 1979), p. 21.

32. "Kitt Peak National Observatory Monthly Report" (April - June 1973), pp. 1-8.

33. "Kitt Peak National Observatory Monthly Report" (November-December 1968), pp. 1-2.

34. Hilliard, R.L. "Steward Observatory's New 90-inch Reflector," *Sky & Telescope*, *34*, 2 (August 1967), pp. 79-81.

35. Scott, John M. *Arizona Highways* (December 1977), pp. 26, 33.

Beneath this panoply of scarlet light, shadows deep and blue creep out from the foothills. Quickly they climb the jagged heights. High on a fortress peak dark shadows clasp the bright red robes of twilight, and fold them in a crevice of the night.

Darkness wraps her mantle of silence around the shoulders of the world. The wizard moon steps onto a mountain top, to orchestrate a soft ballet of moonbeams on a silver lake. Silently, one by one, in the infinite meadows of heaven, blossom the lovely stars, Longfellow's forget-me-nots of the angels.

It is the time of the "long eyes", the astronomers of Kitt Peak National Observatory....For them the coming of darkness ushers in still more wonders....[35]

John M. Scott

Index

Note: Page numbers in **bold** refer to illustrations or illustration captions. Page numbers in *italic* refer to material in direct quotations.

131

M.M. Sundt Construction Co., Tucson, contractor, pier, building, and dome for Nicholas U. Mayall telescope, 87

Monet, Dave, viii

Mount Hamilton, Lick Observatory, 45, 115

Mount Palomar, 66

Mount Palomar Telescope. *See* 200-inch telescope

Mount Wilson Observatories, 4, 86, *86*

Allocation of observing time, 2

Murray J. Shiff Construction Co., contractor, #1 36-inch telescope dome, **39**; 84-inch telescope dome, 55; 50-inch Remote Control Telescope building, 77

National Observatory. *See also* Site selection, National Observatory

Conception of, 1-5

Site selection, **6, 7**, 7, **8**, 9-10, **11, 12**, 12-16, **13, 14**, 27

National Radio Astronomy Observatory, 111, **115**, 116, **116**

National Science Foundation, 68, 101

Advisory Panel for the National Observatory, 4, 23, 69

Astronomical Photoelectric Conference sponsor, 1

Fellows, *86*

National Observatory support, 4-5, 23, 25, 59-60, *69*, 75, 87

Nebulae. *See* name of specific nebula

New York Central Railroad, transport of 84-inch disk, 49

NGC 1976, Orion Nebula, **108**

NGC 2024, Horsehead Nebula, **x, xi**

NGC 4565 Galaxy, in constellation Coma Berenices, **57**

NGC 5194, Whirlpool Galaxy, **103**

NGC 6341, globular cluster, **106**

NGC 6523, Lagoon Nebula, **102**

NGC 6611, Eagle Nebula, **104**

NGC 6720, Ring Nebula, **105**

NGC 6992, Veil Nebula, **107**

Nicholas U. Mayall Telescope, **ii**, 33, **84, 85, 109**, 121, **122, 123**

Construction, 87, **89, 90**, 90-91, **92, 93, 95**

Dedication of, 101, *101*

Design, 87, 91, **94**, 95, **96**, 97

Optics and support, **97**, 97-100, **100**

Photographs taken with, **xi**, **102-108**

Planning, 85-87

Pouring of concrete pier, 87, **88**

Site location, 87

Transport of mirror blank, 99

90-inch telescope, University of Arizona, **109**, 111, 114-115, **114**

NSF. *See* National Science Foundation

Observatories. *See* name of the observatory

Observing conditions, deterioration of, 2

Ohio State University, role in establishing National Observatory, 1, 24

158-inch telescope. *See* Nicholas U. Mayall Telescope

120-inch telescope, Mount Hamilton, 45-46, 87

Orion (constellation), NGC 2024 Horsehead Nebula in, **x, xi**

Orion Nebula, NGC 1976, **108**

Owens of Illinois, contractor, 50-inch Remote Control Telescope, 79

Packard Bell 250 computer, use in Remote Control Telescope, 78

Palomar Observatory, 4, 86, *86*

Papago Indian legend of Kitt Peak, *17*, 17, **18**, 19

Papago Indian Reservation, 19

Papago Indians, *17*, 17, **18**, 19-21, **20**

Papago Indian Tribal Council, 25, *59*, 69, *69*, 113

Land lease negotiations with AURA, Inc., 19-21, **20**, 27

Paulsen, Agnes, viii

133

Westinghouse Electric Corp.,
    Sunnyvale, California,
    contractor, for design
    studies, Nicholas U.
    Mayall telescope, 85
Whirlpool Galaxy, NGC 5194, **103**
White Sands Missile Range, 9
    Aerobee launching, 79, **79**
Whitford, A.E., (quote) *4*
Will, George, viii
Willamette Iron and Steel Co.,
    Portland, Oregon, contractor,
    mechanical parts for 84-inch
    telescope, 52, 55
Willmarth, Daryl, **123**
Yerkes Observatory, 4, 7
Zabiskie, W., 60
Zumwalt, Jimmy, *23, 28*

ABOUT THE AUTHOR:

James E. Kloeppel was born in Sioux City, Iowa, on November 19, 1954. The clean, dark summer nights in Iowa nurtured his interest in astronomy at an early age. After receiving his physics degree from Morningside College, Mr. Kloeppel moved to Tucson where he worked as a Technical Assistant/Large Telescope Operator at Kitt Peak National Observatory. Much of the historical information for this book was gathered while Jim lived at the mountain-based observatory.

Being active in both amateur and professional astronomy, Jim Kloeppel has been a familiar figure at many observatories and planetariums scattered across the United States. During the summer of 1977, Jim and his wife, Darlene, served as volunteers with the National Park Service. Under a program sponsored by the National Park Service, American Astronomical Society, and Smithsonian Institute, the Kloeppels traveled over 12,000 miles presenting astronomy programs to campers in seventeen national parks.

An avid writer, Mr. Kloeppel's articles on science, technology, and history have appeared in such publications as *Research Horizons, Alabama Heritage, Civil War Times, Naval History, Electronics World,* and *Electrical Manufacturing.* His latest book, *Danger Beneath the Waves: A History of the Confederate Submarine H.L. Hunley,* was published in 1987. Mr. Kloeppel is currently employed as a science writer with the Georgia Institute of Technology in Atlanta, Georgia.